Blockchain-based Peer-to-Peer Transactions in Energy Systems

Online at: https://doi.org/10.1088/978-0-7503-6295-5

Blockchain-based Peer-to-Peer Transactions in Energy Systems

Nnamdi Nwulu
Centre for Cyber-Physical Food, Energy and Water Systems (CCP-FEWS), University of Johannesburg, Johannesburg, South Africa

Uyikumhe Damisa
Centre for Cyber-Physical Food, Energy and Water Systems (CCP-FEWS), University of Johannesburg, Johannesburg, South Africa

IOP Publishing, Bristol, UK

ISBN 978-0-7503-6295-5 (ebook)
ISBN 978-0-7503-6293-1 (print)
ISBN 978-0-7503-6296-2 (myPrint)
ISBN 978-0-7503-6294-8 (mobi)

DOI 10.1088/978-0-7503-6295-5

Version: 20231201

IOP ebooks

British Library Cataloguing-in-Publication Data: A catalogue record for this book is available from the British Library.

Published by IOP Publishing, wholly owned by The Institute of Physics, London

IOP Publishing, No.2 The Distillery, Glassfields, Avon Street, Bristol, BS2 0GR, UK

US Office: IOP Publishing, Inc., 190 North Independence Mall West, Suite 601, Philadelphia, PA 19106, USA

Contents

Author biographies

Uyikumhe Damisa

Dr Uyikumhe Damisa is a research scholar at the Centre for Cyber-Physical Food, Energy and Water Systems (CCP-FEWS), University of Johannesburg, South Africa and a postdoctoral fellow in the University. He is a recipient of the Global Excellence Stature Post-doctoral Research Fellowship Award. He has served the academic community in various capacities, including lecturing power systems and supervising energy-related research. His research profile includes research in optimal energy management and blockchain-based P2P transactions in energy systems. He has co-authored journal publications and a book in the fields of renewable energy optimisation and blockchain-based P2P energy transactions.

Nnamdi Nwulu

Professor Nnamdi Nwulu is the Chair Holder of the South Africa/Switzerland Bilateral Research Chair in Blockchain Technology, Professor of Sustainable Cyber-Physical Engineering Systems, and Director of the Centre for Cyber-Physical Food, Energy, and Water Systems (CCP-FEWS) at the University of Johannesburg. Professor Nwulu is also the Acting Head of the School of Electrical Engineering, University of Johannesburg.

Professor Nwulu's research focuses on Food, Energy and Water Systems (FEWS) and the inherent nexus between all three. Specifically, his research focuses on applying FEW-N (Food Energy and Water Nexus) thinking, digital technologies, mathematical optimisation techniques, and machine learning algorithms in food, energy, and water systems. This research focus has produced several books, inventions, and over 200 publications, winning him the 2021 NRF Research Excellence Award for Early Career/Emerging Researchers in Engineering/Technology. He was also a 2020/2021 TW Kambule-NSTF-South32 Award finalist in the Emerging Researcher category and won the 2020 UJ Vice-Chancellors' most promising young researcher award. He is a member of the prestigious South African Young Academy of Science (SAYAS).

He is a professional engineer registered with the Engineering Council of South Africa (ECSA), a senior member of the Institute of Electrical and Electronics Engineers (SMIEEE), a senior member of the South African Institute of Electrical Engineers (SMSAIEE) and a Y-rated researcher by the National Research Foundation of South Africa.

Prof. Nwulu is the editor-in-chief of the *Journal of Digital Food Energy and Water Systems* (JD-FEWS), and associate editor of the *IET Renewable Power Generation* (IET-RPG) and the *African Journal of Science, Technology, Innovation and Development.*

IOP Publishing

Blockchain-based Peer-to-Peer Transactions in Energy Systems

Nnamdi Nwulu and Uyikumhe Damisa

Chapter 1

Introduction to blockchain and smart contracts development

Blockchain is a nascent technology with disruptive and beneficial potentials in a wide variety of industries. This chapter provides a general introduction to the technology whilst touching on the concept of cryptographic hashing, on which the blockchain heavily relies. Blockchains come with varied features and architectures, based on which they can generally be categorised. These categories are discussed in this chapter. As a decentralised technology that is capable of facilitating peer-to-peer (P2P) transactions without a central authority, various consensus mechanisms are employed to reach consensus on the state of the blockchain. Some of the most common mechanisms are presented in this chapter. Some of the key characteristics of blockchains are also enumerated. Various tools and environments have been developed to aid the development of smart contracts and other components of the blockchain ecosystem. These support the development and testing of smart contracts for blockchain-based applications. In addition, with the use of simulated blockchain environments or blockchain test-nets, the cost of deploying and testing smart contracts on a real-world public blockchain is eliminated. In this chapter, with a focus on the Ethereum blockchain, different development tools and environments are presented. The component functions of a sample Ethereum token standard are also discussed. This chapter closes with a discussion of different decentralised application architectures.

1.1 Introduction

We interact with many sectors of society based largely on trust. From the supply chain industry to the pharmaceutical sector, the trust-based model forms the basis for our dealings with various industries. We have to trust that the details stated on the label of the food we consume are true and that the medications we take come from the indicated companies. Financial institutions must be trusted to settle

doi:10.1088/978-0-7503-6295-5ch1

electronic payments, facilitating e-commerce (Nakamoto 2008). Besides the need to trust third parties, extra costs are incurred by the involvement of third parties. Centralisation has also pervaded our society. From financial institutions to the leading technology companies, the centralised model has been utilised. This causes censorship to thrive and unfair exploitation of user data. The need for trust, censorship, and other vices associated with highly centralised systems can be eliminated by using secure decentralised schemes. Blockchain is an emerging technology that has triggered a decentralised revolution (Wang and Su 2020). It is a prominent technology for the enablement of secure peer-to-peer transactions. The success of Bitcoin, a blockchain-based cryptocurrency, has triggered enormous interest in the technology. It is now being explored in various sectors including supply chain (Chang and Chen 2020, Dutta *et al* 2020, Centobelli *et al* 2022), finance (Treleaven *et al* 2017, Gorkhali and Chowdhury 2022, Patel *et al* 2022), healthcare (Hölbl *et al* 2018, Agbo *et al* 2019, McGhin *et al* 2019), energy (Damisa and Nwulu 2022, Damisa, Nwulu *et al* 2022, Damisa, Oluseyi *et al* 2022), voting, real estate (Garcia-Teruel 2020, Gupta *et al* 2020, Saari *et al* 2022), insurance (Gatteschi *et al* 2018, Raikwar *et al* 2018, Popovic *et al* 2020), education (Skiba 2017, Han *et al* 2018, Bhaskar *et al* 2020), and identity management (Dunphy and Petitcolas 2018, Kuperberg 2020, Liu *et al* 2020), among others.

1.2 Blockchain technology

A blockchain is a digital distributed ledger that stores and manages records of all of the transactions ever completed between peers in a network (Afzal *et al* 2022). It can also be seen as a database of transaction records distributed, validated, and maintained by computers spread across a network (Sarmah 2018). The records are compiled into cryptographically and chronologically linked blocks and stored across multiple nodes in the network (Wang *et al* 2019, Gad *et al* 2022). While the full nodes on the network download the entire blockchain and are responsible for validating transactions in a block, the light nodes only verify transactions of interest (Yu *et al* 2020). In addition to the Genesis block, which is the first block of a blockchain network, all blocks contain information about the preceding blocks. The content of a block is cryptographically hashed and stored in the following block, thereby ensuring the immutability of blockchain data. The general structure of a block is shown in figure 1.1. As seen in the figure, the structure of a block is made up of a block header and a block body. While the header stores the previous block hash, a timestamp, among other data, the body stores the transactions captured by the block. Monrat *et al* (2019) describe the block header attributes as follows: the block version determines the rules by which the block should be validated. The previous block hash is a hash pointer of 256 bits that points to the preceding block. The Merkle root is the final hash arrived at after hashing all the transactions on the block. Figure 1.2 shows the concept of the Merkle tree. The timestamp is the time in seconds with reference to 1970–01-01T00:00 UTC. Nonce is a 4-byte field that increases with every hash calculation.

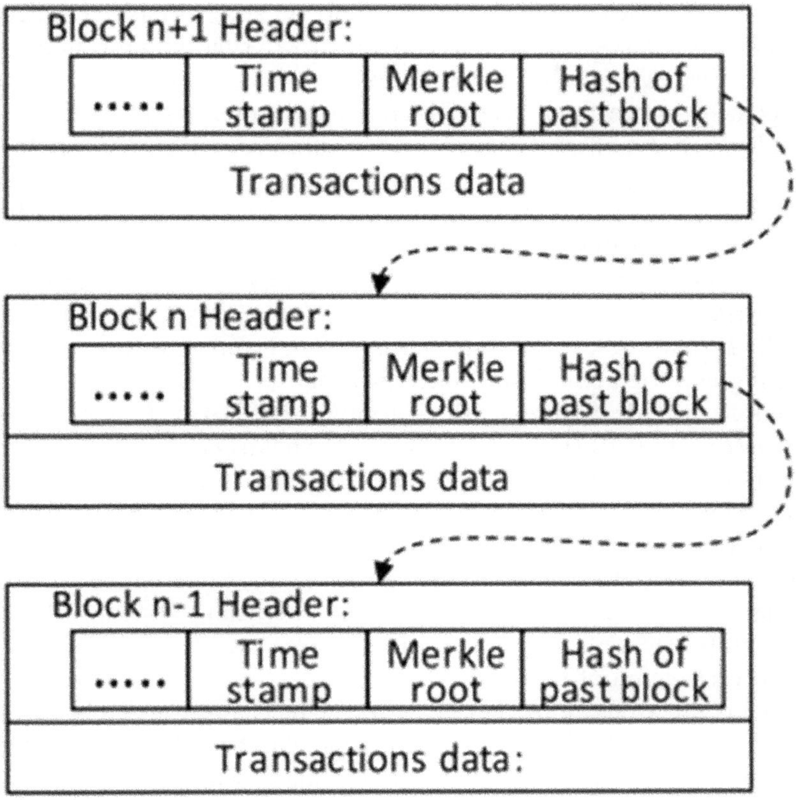

Figure 1.1. Structure of a block Reprinted from Sanka *et al* (2021), Copyright (2021), with permission from Elsevier.

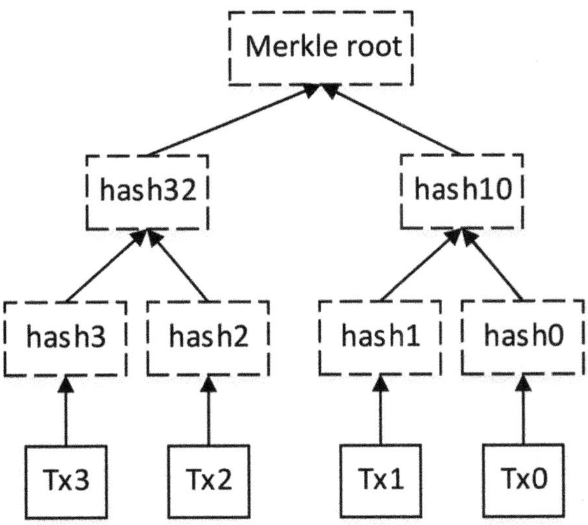

Figure 1.2. The concept of Merkle tree. Reprinted from Sanka *et al* (2021), Copyright (2021), with permission from Elsevier.

The blockchain relies on cryptographic protocols such as hashing functions and digital signatures for its security (Guo and Yu 2022). A hashing function is one that returns a fixed value/length for any given input. It is deterministic in that it always outputs the same value for a given input. Furthermore, a slight alteration in the input results in a significant change in the value returned. A hash function is designed in such a way that the original input cannot be obtained by passing its output through the function. Additionally, a good hash function should be able to compute hashes fast without expending much computing resources (Kaushik *et al* 2017). Hash functions can be categorised as follows: one-way hash functions, collision resistant hash functions, and universal one way hash functions (Sobti *et al* 2012). Digital signatures, time stamping, and authentication can be implemented using cryptographic hash functions (Sobti *et al* 2012). Some examples of hash functions include Secure Hash Algorithm 2 (SHA-2; Guesmi *et al* 2019), Keccak (Mestiri *et al* 2017), and PHOTON (Al-Shatari *et al* 2020).

Transactions within the body of the block are hashed, paired, and hashed again up to the Merkle root, thereby forming a Merkle tree (Singhal *et al* 2018). Computing Merkle roots involves recursive hashing from the bottom, where the transactions lie, up to the final hash that determines the Merkle root (Castellon *et al* 2021). The construction of Merkle trees is a crucial element of blockchain computations (Castellon *et al* 2021). Merkle trees facilitate efficient and secure storage and verification of blockchain data.

Blockchains use an asymmetric cryptography mechanism to authenticate transactions (Monrat *et al* 2019). In a trustless setting, like that of a blockchain network, verifying that a transaction originated from the stated sender is crucial. As a result, network participants own private keys, which are only known to them, and with which they create signatures (Singhal *et al* 2018). Additionally, they also have public keys that are accessible to other network users (Thompson 2017). While the private key is used to encrypt or sign a transaction, the public key is used to decrypt transactions (Monrat *et al* 2019). Figure 1.3 shows the transaction verification process.

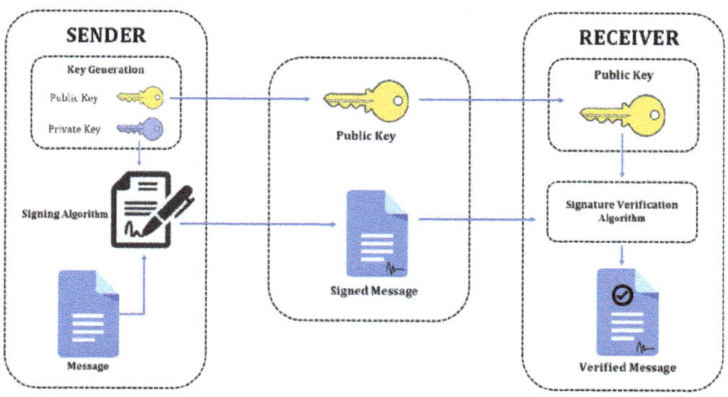

Figure 1.3. Transaction verification process. Reprinted from Latif *et al* (2021), John Wiley & Sons (2021).

1.3 Types of blockchain

Blockchains can be broadly classified into three categories: permissioned, permissionless, and consortium/federated (Sheth and Dattani 2019). A brief description of each type is given below.

1.3.1 Permissioned or private blockchains

These are private blockchains with controllable network and data access. Access to the network is permissioned, i.e., access must be granted by an authority (Polge *et al* 2021). Only authorised nodes can participate in the consensus process and access blockchain data (Peng *et al* 2021). Although few nodes may have the right to read blockchain data, write permission is typically given to a single entity (Olsen *et al* 2022). In addition, transaction verification and block validation can only be performed by authorised users (Sheldon 2019). Permissioned blockchains are typically owned and used by organisations that are able to exercise the necessary controls and access restrictions. Although financial markets and cryptocurrencies tend to employ permissionless blockchains, institutions and businesses lean toward permissioned blockchains (Helliar *et al* 2020). With relatively fewer authorised network participants than those in public blockchains, consensus is reached more quickly in private blockchains, and therefore transaction speeds are higher (Yang *et al* 2020). Furthermore, private blockchains are the preferred option when scalability and data privacy are crucial (Olsen *et al* 2022). However, the nodes that make up the controlling organisation or entity are vulnerable to security breaches not associated with public blockchains (Olsen *et al* 2022).

1.3.2 Permissionless or public blockchains

As the name suggests, these are public blockchains that are generally accessible to the public. Without permission, any interested member of the public can download a copy of the blockchain software, run it on a local computer, and be eligible to participate in the consensus process (Olsen *et al* 2022). Anyone can submit or confirm transaction and participate in the transactions validation process (Solat *et al* 2021). A digital infrastructure is permissionless if, without obtaining permission from an authority, it can be used, developed, and governed by obeying publicly declared protocols (Nabben and Zargham 2022). In public blockchains, user-initiated blockchain state changes, which are referred to as transactions, are cryptographically signed, broadcast on the network, and verified (Neudecker and Hartenstein 2018). With the help of a block explorer, these transactions are visible to the public. These transactions public blockchains suggest more decentralised architecture than private blockchains (Neudecker and Hartenstein 2018). In fact, Olsen *et al* (2022) categorically regard private blockchains as a centralised network, considering that a single entity controls it. Examples of public blockchains include Bitcoin, Ethereum, Cardano, Dogecoin, and Litecoin, among others.

1.3.3 Federated or consortium blockchains

Consortium blockchains are neither absolutely permissioned or permissionless. Unlike public blockchains, which allow anyone to participate and verify blocks in the system, consortium blockchains rely on an authorised group of parties to validate blocks (An and Chen 2021). Unlike private blockchains, which are operated by single entities, a set of entities typically manage consortium blockchains (Dib *et al* 2018, Zhang and Lin 2018); specific nodes are authorised to maintain the shared database (Li *et al* 2017). However, consortium blockchains leverage the privacy and efficiency of private blockchains while benefiting from the decentralised governance of public blockchains (Dib *et al* 2018). They can be used to facilitate efficient inter-organisational collaboration and transactions. The banking sector typically uses consortium blockchains (Olsen *et al* 2022). They also offer a high level of security (Ashfaq *et al* 2022). Regarding their architecture, consortium blockchains are more decentralised than private blockchains (Dib *et al* 2018). Consortium blockchains have the potential to facilitate document handling, lessen data redundancies, replace legacy systems, and reduce transaction costs (Olsen *et al* 2022).

1.4 Consensus mechanisms

Unlike traditional centralised systems, where all of the data is stored in central databases, a blockchain operates a distributed ledger system with data being held across multiple nodes. To ensure that all nodes agree on the state of the blockchain, i.e., the record of transactions in the ledger, consensus mechanisms are used (Bouraga 2021). A consensus algorithm helps to develop an accurate ledger shared in a network among nodes without a central authority (Omote and Yano 2020, S. Zhang and Lee 2020). The nodes validate the data to be included in the blockchain and agree on the order in which they are included using a consensus mechanism (Cachin and Vukolić 2017); an agreement must be reached on the blocks to be added to the chain (Bouraga 2021). Consensus algorithms also enable unanimous decision making in decentralised networks (Sankar *et al* 2017). They are crucial to ensuring the integrity and security of blockchain networks (Nguyen *et al* 2019). The choice of consensus mechanisms significantly impacts the fault tolerance, transaction capacity, and scalability of a blockchain network (Xiao *et al* 2020). Many types of consensus mechanism have been developed. The most common are Proof-of-Work (PoW), Proof-of-Stake (PoS), Delegated Proof-of-Stake (DPoS), and Practical Byzantine Fault Tolerance (PBFT).

1.4.1 Proof of work

The PoW mechanism is the earliest consensus protocol and is used by the most prominent cryptocurrency—Bitcoin (Schinckus 2021). In a blockchain that utilises the PoW consensus mechanism, certain nodes, called miners, compete to append the next block to the chain. The competition involves solving a problem using a procedure that consumes a lot of computational resources. The chances of winning the competition and mining the block increase with the amount of computing

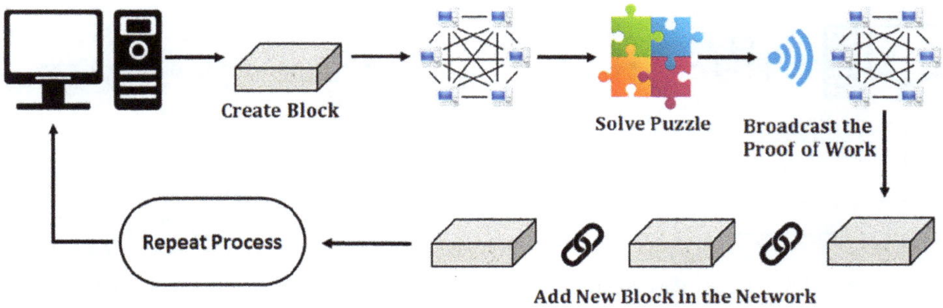

Figure 1.4. Flow of PoW. Reprinted from Latif *et al* (2021) John Wiley & Sons (2021).

resources available to a miner, and miners are rewarded with some native coins for each block that they successfully mine (Vashchuk and Shuwar 2018). The competition involves the repeated hashing of a nonce, previous block hash, etc, while changing the nonce after each iteration until a set target is met (Zhang and Lee 2020). The set target is the number of leading zeros that must be contained in the resulting hash output (Bouraga 2021). Apparently, the solution to this PoW's computational problem cannot be reached by applying mathematical principles or any form of logic, but may be reached by trial and error. Although solving the problem typically involves a high consumption of computing resources, the solution can easily be verified by other network nodes. Upon finding the nonce that hashes to a value within the target range, the miner broadcasts the new block together with its transactions, other nodes verify it, and it is added to the chain of blocks (Nguyen *et al* 2019). While it is possible for a malicious node to overthrow a new blockchain with only one or few blocks, a long chain will require a considerable amount of computational power for a successful attack. (Zhang and Lee 2020). As the blockchain network grows, so does the complexity of the computational problem that needs to be solved by prospective miners (Hazari and Mahmoud 2019). This complexity is reflected in the number of leading zeros that the output of the hashed block header must have, which increases with the number of zeros required (Bouraga 2021). Figure 1.4 shows the flow of the PoW consensus mechanism.

1.4.2 Proof of stake

The PoS consensus mechanism is an alternative to PoW, which requires less computational resources. Block miners are called validators in the PoS protocol. Validators are required to solve a mathematical puzzle, i.e., it has the block header to meet a certain target. However, the energy expended in the process is not as much as is used in PoW (Zhang and Lee 2020, Bouraga 2021). To be eligible to validate a block of transactions, nodes are required to stake some of the network's native token. The staked tokens are locked up and the winning validator is rewarded with network transaction fees (Saad and Radzi 2020). A node's chances of validating a new block increase with the amount of tokens staked (Bouraga 2021). This ensures that the network's security is largely in the hands of those with the greatest value at

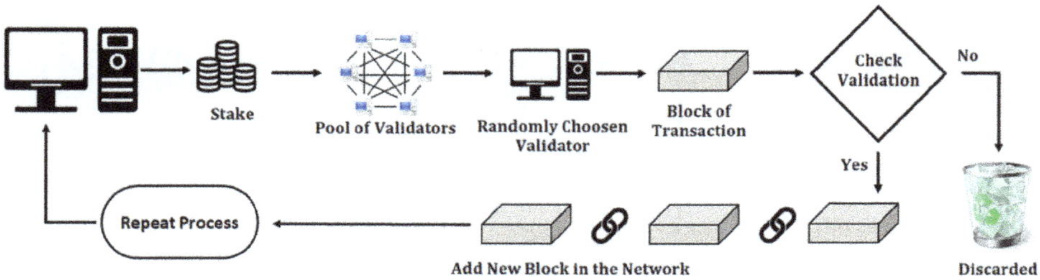

Figure 1.5. Flow of PoS. Reprinted from Latif *et al* (2021) John Wiley & Sons. (2021).

stake if the network is compromised. For a PoS-based blockchain network to be compromised, the attacker(s) must acquire most of the network's native token, which can be very expensive, especially for large, established networks (Vashchuk and Shuwar 2018). Figure 1.5 shows the flow of PoS.

1.4.3 Delegated proof of stake

The key difference between the PoS and DPoS consensus mechanisms lies in the approach used to select block validators (Cheng *et al* 2019). In DPoS, the potential block verifiers are elected by the nodes that hold stakes in the network to participate in the consensus process and produce the blocks in turn (Hu *et al* 2021). The reward for block creation is shared among the elected delegate and the nodes that elected it (Saad and Radzi 2020). Delegates who do not produce a block in their turn are replaced (Zhang and Lee 2020). Since DPoS involves no mining process, the nodes involved in block verification and production are relatively fewer and the time to reach consensus is reduced (Cheng *et al* 2019). However, while the DPoS protocol improves speed and throughput, it introduces security vulnerabilities (Cheng *et al* 2019, Saad and Radzi 2020, Hu *et al* 2021).

1.4.4 Practical Byzantine Fault Tolerance

PBFT (Castro and OsDI 1999, Lamport *et al* 2019, Wang *et al* 2021) is a consensus mechanism which suits a scenario where a number of Byzantine generals who are physically distant apart need to reach a consensus, by communicating using messengers, regarding the action to be taken against an enemy. PBFT offers a general solution to the consistency issue in distributed systems, while Byzantine error nodes are present (Cheng *et al* 2019). In a PBFT-based blockchain network, communication occurs between nodes, through the network, to reach a consensus on the blocks to be appended. Five phases are involved in the process: Request, Pre-prepare, Prepare, Commit, and Reply (Zhang and Lee 2020). In summary, a client sends a proposed block to a primary node, which in turn broadcasts the block to several other nodes. The proposed block is accepted into the chain if a certain number of nodes agree to it (Zhang and Lee 2020, Bouraga 2021). The implementation of PBFT ensures that in every round of consensus, nodes act consistently and maintain the same state of the system (Zhang and Lee 2020).

In addition to the consensus mechanisms, other mechanisms include Proof of Elapsed tTme (Chen *et al* 2017), Proof of Capacity (Aggarwal and Kumar 2021), Leased PoS (Alyaseen 2019), Proof of Authority (Manolache *et al* 2022), Scalable Byzantine Consensus Protocol (Berger and Reiser 2018), Proof of Bandwidth (Cong and Zi 2020), Proof of Burn (Aggarwal and Kumar 2021), and Proof of Importance (Krishnamurthi and Shree 2019).

1.5 Key characteristics of blockchains

1.5.1 Decentralisation

Unlike traditional systems, which heavily depend on central authorities and servers, blockchain network nodes cooperate to keep the network running and secure without a central authority. Multiple nodes are engaged in a distributed consensus mechanism, agree on the state of the blockchain, and each store it. Hence, they have access to the entire history of transactions on the blockchain (Tapscott and Tapscott 2017). Therefore, blockchains do not provide a single point of failure (Nguyen *et al* 2019). Reduced reliance on servers in blockchain-based systems can result in improved system performance and reduced server-related costs (Olsen *et al* 2022).

1.5.2 Immutability

Cryptographic mechanisms are used to effectively chain all of the blocks of transactions, from the genesis block to the current block. In other words, each block represents all of the transactions recorded in all the preceding blocks. With the help of computational algorithms and protocols, the records are chronologically ordered and permanent (Tapscott and Tapscott 2017). These blocks of transactions are stored and can be validated by multiple network nodes, making it nearly impossible to falsify them (Olsen *et al* 2022). They are generally tamper proof, i.e., very difficult to modify (Nguyen *et al* 2019).

1.5.3 Auditability

The movement of assets between nodes of a blockchain network is immutably recorded in a distributed ledger, beginning from the genesis block to the most recent block of transactions. This facilitates the auditability of blockchain-based systems.

1.6 Smart contract development tools

Smart contracts are pieces of code that are stored and executed on a blockchain. Figure 1.6 shows a simple, smart contract mechanism. The blockchain development ecosystem has attracted many developers around the world. As a result, many development communities have emerged. To lower the barrier of entry for new entrants into the blockchain space, these developers have created environments and tools to facilitate the development and testing of the various components that make up the blockchain ecosystem. A few of these tools are highlighted and briefly described below.

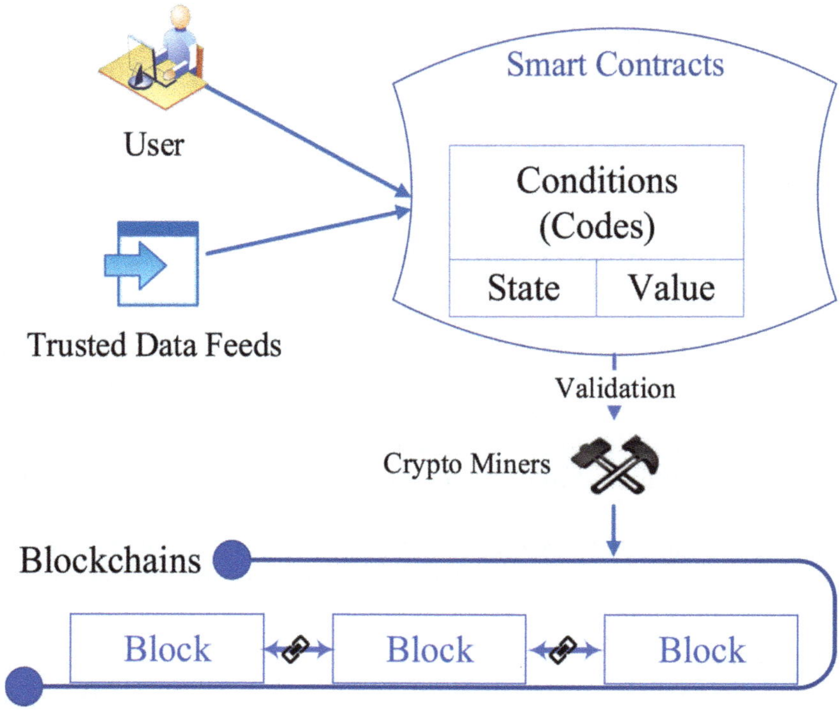

Figure 1.6. Smart contract mechanism. Reprinted from Sookhak *et al* (2021), Copyright (2021), with permission from Elsevier.

1.6.1 Remix IDE

The Remix Integrated Development Environment is a tool that is intended for smart contract developers of all levels of expertise. The environment offers support throughout the smart contract creation journey, from creation to testing and deployment. The Remix integrated development environment (IDE) comes in different flavours. Its web-based flavour can be used directly on the web without the need to setup an environment or install additional software, while the Remix desktop application can be used on a PC. It also has a Visual Studio Code extension that makes it possible to write smart contracts in Visual Studio Code. Smart contracts do not operate as a standalone entity, they need to be deployed to a blockchain; hence, the Remix IDE gives developers different deployment environments. One such option is the Remix virtual machine (VM), which is a simulated blockchain that runs in the browser window. It provides multiple nodes with which you can interact with the smart contract. The nodes are preloaded with virtual cryptocurrency that can be used for transaction fees or transferred by a smart contract between nodes. Smart contracts can also be deployed in an Injected Provider MetaMask environment. This option requires MetaMask to be embedded in the browser. Other deployment environments are available within the Remix IDE. Remix's functionality can be extended by enabling plugins provided in the environment. Although Remix offers some functionalities crucial to the successful development and deployment of smart contracts, as well as a user-friendly interface suitable for beginners, it has limitations. It is, however, a more flexible alternative in the Truffle framework.

1.6.2 Hardhat

The Hardhat development environment comprises various components that can be used to edit, compile, debug, and deploy Ethereum-based smart contracts and decentralised applications. Recurring tasks associated with developing these applications are managed and automated using Hardhat's main component, Hardhat Runner. At its core, these services entail constructing, executing, and testing smart contracts (Swati *et al* 2022).

1.6.3 Truffle

Truffle is a development environment that supports the creation and testing of smart contracts intended for blockchains that use the Ethereum virtual machine (EVM), which is a popular framework within the Ethereum development community. It is more easily accessible to experienced developers because it operates a command-line interface. Truffle can be run on Windows, Linux, or Mac operating systems. It requires Node.js, which can be installed using the Node Package Manager (NPM). With the same package manager, the Truffle software can be installed. After installation, Truffle can be used by first creating a new project directory and then initialising a Truffle project within the directory, as follows:

```
uyi@uyi:~$ mkdir UJToken
uyi@uyi:~$ cd UJToken
While in the newly created folder, a new Truffle project is initialised
uyi@uyi:~/UJToken$ truffle init
Starting init...
================
Copying project files to /home/uyi/UJToken
Init successful, sweet!
The content of the folder can be viewed using the ls command.
uyi@uyi:~/UJToken$ ls
contracts migrations test truffle-config.js
```

The initialisation of a new Truffle project creates new directories in the project folder. One of these is the contracts directory, which holds all the solidity contracts of the project. At initialisation, by default, a Migrations.sol contract, which must be deployed to utilise the migrations feature offered by Truffle (https://www.trufflesuite.com/docs/truffle/getting-started/running-migrations#migration-files [03/02/2021]).

```
uyi@uyi:~/UJToken$ ls contracts
Migrations.sol
```

A new Truffle project also comes with a migrations directory, which manages contract deployment. By default, it holds a javascript file that manages the deployment of the Migrations.sol contract. Subsequent smart contracts will require the creation of additional JavaScript files for the management of their deployment.

```
uyi@uyi:~/UJToken$ ls migrations
1_initial_migration.js
```

Further details on the use of Truffle can be found on the product's website: trufflesuite.com

1.6.4 OpenZeppelin

Developers of a nascent technology such as smart contract typically have only a few years of experience in the field. Moreover, many blockchain development tools have quite a demanding learning curve. Expectedly, some deployed smart contracts have been discovered to have bugs, some of which have been exploited by hackers, resulting in the loss of assets. Smart contracts can hold assets of very high value, and are therefore targeted by malicious attackers. OpenZeppelin aids the building, automation, and operation of decentralised applications with its security products. It offers a collection of community-vetted, modular, reusable, secure Ethereum smart contracts, and a platform that automates Ethereum operations and facilitates quick delivery of high-quality products. The modular form of smart contracts offers developers the flexibility to adapt them to their specific needs.

The following illustrates a simple ERC-20 token based on the Openzeppelin implementation:

The library containing secure modular contracts from OpenZeppelin can be downloaded using the Node Package Manager (npm) with the command:

```
$ npm install @openzeppelin/contracts
```

This package can then be used flexibly to develop customised smart contracts via contract inheritance. Of the various contract variants provided, we select the ERC-20 variant. This variant implements all the mandatory functions specified in the ERC-20.

```
pragma solidity ^0.6.0;
import '@openzeppelin/contracts/token/ERC20/ERC20.sol';
contract UJToken is ERC20
constructor(uint256 initialSupply) ERC20('UJToken', 'UJT')
_mint(msg.sender, initialSupply); }}
```

The first line of the code above specifies the version of Solidity for proper code compilation. The caret symbol before the version number implies that a compiler of version 0.6.0 or greater should be used. The next line imports the particular variant of the token contract that we would be inheriting from. Our contract begins on the third line with the Solidity keyword contract and an identifier, UJToken. The keyword depicts inheritance; hence UJToken inherits the particular Openzeppelin variant implementation identified as 'ERC-20'. The details of this implementation can be viewed in OpenZeppelin's Github repository (https://github.com/OpenZeppelin/openzeppelin-contracts/blob/master/contracts/token/ERC20/ERC20.sol 28/01/2021). When a contract inherits from another, the inherited contract is referred to as the child, while the one inherited is the parent.

The constructor keyword is used to declare a function that is invoked only at the point of deployment of the smart contract. Constructors are optional in smart contracts; however, in the absence of a constructor, a default constructor is used. In our implementation, the constructor takes the initialSupply argument, which is an unsigned integer. By implication, an unsigned integer representing the amount of tokens to be generated would have to be passed into the constructor before contract creation. The Openzeppelin ERC-20 contract also has its constructor, which also requires some arguments to be passed, i.e., token name and token symbol. When a contract inherits from another contract, the required arguments of the constructor of the parent contract must be supplied. It should be noted that it is possible to inherit from multiple contracts simultaneously, in which case the arguments of all the constructors of the parent contracts must be supplied. As we noted earlier, constructors are optional—parent contracts without constructors need not receive any argument. The body of our constructor invokes a function _mint, which is contained in the parent contract—ERC-20. The _mint function is designed to create new tokens and allocate them to a specified address. It receives two arguments, msg.sender and initialSupply. Meanwhile, msg.sender returns the address of the caller of a function; however, since it is to be executed within the body of a constructor, it will return the address of the deployer of the contract.

The mint function therefore creates initialSupply tokens and assigns them to msg.sender. A token contract has therefore, been created with a specified amount of tokens, a token name, and its symbol. So far, we have only successfully written the code for the token smart contract. The next step would be to compile the code because Solidity is a compiled programming language. The code gets compiled into a language that the EVM can understand, i.e., Bytecode. Remix is an on-line, in-browser IDE where smart contracts can be easily compiled. The following URL https://remix.ethereum.org/#optimize=false&runs=200&evmVersion=null&version=soljson-v0.7.4+commit.3f05b770.js (29/01/2021) leads to the environment.

1.6.5 Ganache

Ganache is a simulated personal blockchain that can be run on a local computer. It facilitates rapid distributed application development for Filecoin and Ethereum. It offers a deterministic and secure environment for the development, deployment, and testing of decentralised applications (dApps). Ganache has a user interface (UI) and a command-line interface (CLI). The simulated local blockchain offers multiple accounts preloaded with virtual cryptocurrency that can be transferred between accounts/addresses/nodes or used to pay for transaction fees.

1.7 Sample smart contract: ERC-20 token standard

The Ethereum blockchain extends the functionality of Bitcoin through its support for smart contracts. With these contracts, tokens (or cryptocurrencies) other than the Ethereum native token, Ether, can be built, held, sent, and received on the Ethereum chain. There are a vast number of things that such tokens can represent, e.g., they can represent shared ownership of a tangible asset, scores earned in a game, financial assets, or fiat currency, among many others. The ERC-20 standard regulates the development of fungible tokens on the Ethereum blockchain. Fungible tokens are nondistinct tokens that are equal in value, similar to fiat currencies. For this kind of token, the value corresponds to the quantity held. Meanwhile, nonfungible tokens refer to unique tokens that represent specific assets. Their value is tied to the distinct assets that they represent. The ERC-20 token standard states the required functionalities for ERC-20-compliant tokens. As a result of this standard, decentralised token exchange platforms are able to interact with these tokens seamlessly. Tokens can be bought with Ethers or exchanged with other tokens. They can also be awarded to incentivise some good behaviour. An ERC-20 token must have the following functions (https://eips.ethereum.org/EIPS/eip-20):

totalSupply

The totalSupply function is required and expected to be used to view the total number of the token in circulation in real-time. The total supply of tokens includes those held at the contract's address as well as those in other accounts. In Solidity, it is defined as:

```
The function totalSupply() public view returns (uint256)
// function logic}
```

The first word function is a Solidity keyword that is used to define a function. The next word function, totalSupply, identifies the function. The following empty parentheses show that the function takes in no arguments or has no parameters. The public keyword makes the function accessible to anyone interested, and the view keyword restricts the function to a read-only function. It only reads the state of the blockchain and cannot alter it. The returns keyword specifies in the following parentheses the type of value that the function is expected to return. In Solidity, uint256 stands for an unsigned integer with 256 bits. The curly braces provide the section to input the function's logic, i.e., all the function is expected to do when it is called.

balanceOf

The balanceOf function is another required and publicly-accessible function that returns the number of tokens held in the account identified as address _owner.

```
function balanceOf (address _owner) public view returns (uint256 balance) {
// function logic}
```

The caller of this function would be required to pass in an argument of type address, which is the variable type used to store account/contract addresses, with the identifier _owner. The function returns an unsigned integer variable balance. Depending on the logic specified in the body of the function, access to account balance may be restricted to the owner of the account only.

transfer

The transfer function enables the transfer of tokens between accounts. The address to which the token should be sent, _to, and the number of tokens to be sent, _value, are passed as arguments to the function. The tokens are transferred from the caller's address of the function. As a security measure, a check must be performed to ensure that the balance of the caller's account is at least equal to that being transferred. Upon successful completion of the task, it returns a Boolean value.

```
function transfer(address _to, uint256 _value) public returns (bool success){
// function logic}
```

transferFrom

The transferFrom enables the transfer of tokens from one account to another. The addresses of both accounts and the number of tokens to be transferred are passed in as arguments to the function. The body of the function must contain a necessary check that the caller of the function has permission to do the transfer. Upon successful completion of the task, it returns a Boolean value.

```
function transferFrom(address _from, address _to, uint256 _value) public
   returns (bool success){// function logic}
```

approve

The approve function allows an account holder to grant permission to another account holder to withdraw up to a certain amount of tokens from the former's account. The address of the account being granted such permission and the amount being permitted, are passed in as arguments to the function. It returns a Boolean value to indicate the success of the operation.

```
function approve(address _spender, uint256 _value) public returns (bool suc-
   cess){
// function logic}
```

allowance

The allowance function is used to view the amount left of that which was approved by an account holder to be spent by another. The address of the account's original owner and that of the account being granted permission to spend are passed to the function as arguments. The function then returns the remaining amount.

```
function allowance(address _owner, address _spender) public view returns
   (uint256 remaining){
// function logic }
```

Some optional functions specified by the ERC-20 token standard are as follows:

name

The name function is a publicly-accessible function that can be used to view the name assigned to the token.

```
function name() public view returns (string){
// function logic}
```

symbol

The symbol function is a publicly-accessible function that can be used to view the symbol assigned to the token.

```
function symbol() public view returns (string){
// function logic}
```

decimals

The decimals function is a publicly-accessible function that can be used to view the decimals being used by the token. The decimal value determines the mathematical relationship between the token base unit used in smart contract operations and the user representation of the token (what the user sees on the frontend). For instance, a decimal value of 18 would imply that the amount of the token would be divided by 1000 000 000 000 000 000 to obtain the user representation of the token.

```
function decimals() public view returns (uint8)
// function logic
}
```

1.8 Decentralised applications

Generally, dApps are generally blockchain-based applications that, unlike traditional applications, have some or all of their components running on a blockchain. The critical components of Dapps can be distributed across the nodes of a decentralised blockchain network (Taş and Tanröver 2019). Their integration with blockchain technology introduces features such as immutability, transparency, ownership, and security, which makes them superior to traditional applications in contexts where these qualities are essential (Banerjee 2023). Generally, the key components of a Dapp are a web application, distributed business logic, distributed data, and client-side encryption (Zheng *et al* 2021). The Web application is typically the user-facing component of a Dapp, which provides a connection to the business logic of the Dapp. This connection to business logic may be achieved directly using a blockchain node or indirectly via a centralised server provided by the Dapp provider (Wessling and Gruhn 2018). The web application offers a friendly interface for non-technically inclined users to use Dapps. The distributed business logic is typically implemented by smart contracts, which are executed on the blockchain. Storing all Dapp-related data on-chain is usually inefficient; hence, some data is typically stored off-chain. To ensure a fully decentralised system, such off-chain data could also be stored using decentralised file storage schemes, in which case both on-chain and off-chain data are distributed data. Client encryption is required to authenticate on-chain transactions. It can be achieved by using a Web3 wallet integrated with the browser. Many Dapps have been developed on several real-world blockchain networks. These Dapps adopt varying architectures, which researchers have attempted to categorise. The authors of Kina-Kina *et al* (2019) classified Dapps into two architectural categories. In contrast, the authors of Wessling and Gruhn (2018) surveyed various existing domain-based applications. The first pattern is one in which users can generate and send transactions via direct interaction with smart contract. The second pattern involves users communicating with the Dapp using a crytobrowser or a crypto wallet like MetaMask. In the last identified pattern, users communicate with the Dapp through a website created by the Dapp provider. Wu *et al* (2021) identified three Ethereum Dapp architectures:

A direct architecture, in which users interact directly with deployed smart contracts.
An indirect architecture, in which users interact with smart contracts via a centralised backend server.
A mixed architecture, in which users can interact with smart contracts both directly and indirectly via a centralised backend server.

The authors of Zheng *et al* (2023) categorise Dapps into four classifications:

1. Native client as a Dapp: The user in this architecture uses an installed blockchain client to interact with the network. Bitcoin is a prominent example of such an architecture.
2. Smart contract as a Dapp: The user uses a smart contract browser like Remix to interact with smart contracts.
3. Web and contract as a Dapp: In this widely-used architecture, a web interface is introduced through which users are able to interact with the Dapp. Cryptocurrency wallets can also be used along with the website.
4. Fully decentralised Dapp (Gerring 2014): This proposed architecture decentralises all of the components of a Dapp. In addition to the already decentralised smart contract technology for business logic, Swarm (https://www.ethswarm.org/) can be used for decentralised file storage and Whisper (https://whispermsg.com/) can be used for decentralised messaging.

1.9 Summary

Blockchain possesses a huge potential for disruptive but beneficial transformations in a wide range of industries. The financial sector has experienced the most significant paradigm shift in operation and business models compared to other industries,. The impact of Blockchain has also been felt in supply chain and energy, among other sectors. Depending on the requirements of an application, blockchain offers a varied set of features. A private blockchain, for instance, is more suitable for use by a single organisation because it can exercise more control and access restrictions. On the one hand, a global cryptocurrency is better suited for a public blockchain network. Hence, the appropriate type of blockchain must be selected after carefully considering the key requirements of an application. Consensus mechanisms are crucial to the decentralised operation of blockchain. They typically have strengths and weaknesses. The PoW mechanism, for instance, is highly secure but expends a lot of energy. On the other hand, the PoS mechanism consumes a significantly reduced energy, but is not as secure as PoW. Scalability and transaction speed are other metrics that can be used to assess the performance of consensus mechanisms. The drive to improve certain of these metrics has led to the creation of multiple other consensus mechanisms. A smart contract deployed to a public blockchain is immutable; hence, it is impossible to fix bugs after deployment. It also becomes expensive to test its functionalities. There are tools and environments that can offer a way to test smart contracts before they are deployed to real-world blockchains. Due to their ability to hold or access assets, smart contracts are sometimes targeted by hackers, who attempt to exploit bugs in the system. A package such as Openzeppelin offers various community-vetted smart contracts that can be adapted to specific needs. This helps developers to create more secure smart contracts.

References

Afzal M, Li J, Amin W, Huang Q, Umer K, Ahmad S A, Ahmad F and Raza A 2022 Role of blockchain technology in transactive energy market: a review *Sustain. Energy Technol. Assess.* **53** 102646

Agbo C C, Mahmoud Q H and Eklund J M 2019 Blockchain technology in healthcare: a systematic review *Healthcare* **7** 56

Aggarwal S and Kumar N 2021 Cryptographic consensus mechanisms *Advances in Computers* **vol 121** pp 211–26 (Amsterdam: Elsevier)

Al-Shatari M O A, Hussin F A, Aziz A A, Witjaksono G and Tran X T 2020 FPGA-based lightweight hardware architecture of the PHOTON hash function for IoT Edge devices *IEEE Access* **8** 207610–8

Alyaseen I F T 2019 Consensus algorithms blockchain: a comparative study *Int. J. Cogn. Comput.* **5** 66–71

An H and Chen J 2021 ElearnChain: a privacy-preserving consortium blockchain system for e-learning educational records *J. Inf. Secur. Appl.* **63** 103013

Ashfaq T *et al* 2022 An efficient and secure energy trading approach with machine learning technique and consortium blockchain *Sensors* **22** 7263

Banerjee A 2023 A systematic study on the development of decentralized application (dApp) *Distributed Computing to Blockchain* (Elsevier) pp 173–88

Berger C and Reiser H P 2018 Scaling byzantine consensus: a broad analysis *Proceedings of the 2nd Workshop on Scalable and Resilient Infrastructures for Distributed Ledgers* 13–8

Bhaskar P, Tiwari C K and Joshi A 2020 Blockchain in education management: present and future applications *Interact. Technol. Smart Educ.* **18** 1–17

Bouraga S 2021 A taxonomy of blockchain consensus protocols: a survey and classification framework *Expert Syst. Appl.* **168** 114384

Cachin C and Vukolić M 2017 Blockchain consensus protocols in the wild ArXiv Preprint ArXiv:1707.01873

Castellon C, Roy S, Kreidl P, Dutta A and Bölöni L 2021 Energy efficient merkle trees for blockchains *2021 IEEE 20th Int. Conf. on Trust, Security and Privacy in Computing and Communications (TrustCom)* pp 1093–9

Castro M and OsDI B L 1999 Practical byzantine fault tolerance *Usenix.Org* (https://usenix.org/legacy/publications/library/proceedings/osdi99/full_papers/castro/castro.ps) (accessed 2 February 2023).

Centobelli P, Cerchione R, Vecchio P D, Oropallo E and Secundo G 2022 Blockchain technology for bridging trust, traceability and transparency in circular supply chain *Inform. Manag.* **59** 103508

Chang S E and Chen Y 2020 When blockchain meets supply chain: a systematic literature review on current development and potential applications *IEEE Access* **8** 62478–94

Chen L, Xu L, Shah N, Gao Z, Lu Y and Shi W 2017 On security analysis of proof-of-elapsed-time (poet) *Stabilization, Safety, and Security of Distributed Systems: 19th Int. Symp., SSS 2017 (Boston, MA, November 5–8, 2017)* pp 282–97

Cheng Y, Hu X and Zhang J 2019 An improved scheme of proof-of-stake consensus mechanism2019 *4th Int. Conf. on Mechanical, Control and Computer Engineering (ICMCCE)* pp 826–3

Cong X and Zi L 2020 Blockchain applications, challenges and evaluation: a survey *Discrete Math., Algorithms Appl.* **12** 2030001

Damisa U and Nwulu N I 2022 Blockchain-based auctioning for energy storage sharing in a smart community *Energies* **15** 1954

Damisa U, Nwulu N I and Siano P 2022 Towards blockchain-based energy trading: a smart contract implementation of energy double auction and spinning reserve trading *Energies,* **15** 4084

Damisa U, Oluseyi P O and Nwulu N I 2022 Blockchain-based gas auctioning coupled with a novel economic dispatch formulation for gas-deficient thermal plants *Energie* **15** 5155

Dib O, Brousmiche K-L, Durand A, Thea E and Hamida E B 2018 Consortium blockchains: overview, applications and challenges *Int. J. Adv. Telecommun.* **11** 51–64

Dunphy P and Petitcolas F A P 2018 A first look at identity management schemes on the blockchain *IEEE Secur. Privacy* **16** 20–9

Dutta P, Choi T M, Somani S and Butala R 2020 Blockchain technology in supply chain operations: applications, challenges and research opportunities *Transp. Res. Part E: Logist. Transp. Rev.* **142** 102067

Gad A G, Mosa D T, Abualigah L and Abohany A A 2022 Emerging trends in blockchain technology and applications: a review and outlook *J. King Saud Univ.—Comput. Inform. Sci.* **34** 6719–42

Garcia-Teruel R M 2020 Legal challenges and opportunities of blockchain technology in the real estate sector *J. Prop., Plann. Environ. Law* **12** 129–45

Gatteschi V, Lamberti F, Demartini C, Pranteda C and Santamaría V 2018 Blockchain and smart contracts for insurance: is the technology mature enough? *Future Internet* **10** 20

Gerring T 2014 Building the Decentralized Web 3.0. (Accessed 12 November 2017)

Gorkhali A and Chowdhury R 2022 Blockchain and the evolving financial market: a literature review *J. Ind. Integr. Manag.* **7** 47–81

Guesmi R *et al* 2019 A novel chaos-based image encryption using DNA sequence operation and secure hash algorithm SHA-2 *Nonlinear Dynamics* **83** 1123–36

Guo H and Yu X 2022 A survey on blockchain technology and its security *Blockchain: Res. Appl.* **3** 100067

Gupta A, Rathod J, Patel D, Bothra J, Shanbhag S and Bhalerao T 2020 Tokenization of real estate using blockchain technology *Lecture Notes in Computer Science (Including Subseries*

Lecture Notes in Artificial Intelligence and Lecture Notes in Bioinformatics), 12418 LNCS 77–90
Han M, Wu D, Li Z, Xie Y, He J S and Baba A 2018 A novel blockchain-based education records verification solution *SIGITE 2018—Proc. of the 19th Annual SIG Conf. on Information Technology Education* 178–83
Hazari S S and Mahmoud Q H 2019 A parallel proof of work to improve transaction speed and scalability in blockchain systems *2019 IEEE 9th Annual Computing and Communication Workshop and Conf. (CCWC)* pp 916–21
Helliar C V, Crawford L, Rocca L, Teodori C and Veneziani M 2020 Permissionless and permissioned blockchain diffusion *Int. J. Inf. Manage.* **54** 102136
Hölbl M, Kompara M, Kamišalić A and Zlatolas L N 2018 A systematic review of the use of blockchain in healthcare *Symmetry 2018* **10** 470
Hu Q, Yan B, Han Y and Yu J 2021 An improved delegated proof of stake consensus algorithm *Procedia Comput. Sci.* **187** 341–6
Kaushik A, Choudhary A, Ektare C, Thomas D and Akram S 2017 Blockchain—literature survey *2017 2nd IEEE Int. Conf. on Recent Trends in Electronics, Information and Communication Technology (RTEICT)* 2145–8
Kina-Kina K M, Cutipa-Arias H E and Shiguihara-Juárez 2019 A comparison of performance between fully and partially decentralized applications *2019 IEEE XXVI Int. Conf. on Electronics, Electrical Engineering and Computing (INTERCON)* pp 1–4
Krishnamurthi R and Shree T 2019 A brief analysis of blockchain algorithms and its challenges *Architectures and Frameworks for Developing and Applying Blockchain Technology* (Hershey, PA: IGI Global) 69–85
Kuperberg M 2020 Blockchain-based identity management: a survey from the enterprise and ecosystem perspective *IEEE Trans. Eng. Manage.* **67** 1008–27
Lamport L, Shostak R and Pease M 2019 Concurrency: the works of leslie lamport *Association for Computing Machinery* (New York: ACM) 203–26
Latif S, Idrees Z, Huma Z and Ahmad J 2021 Blockchain technology for the industrial Internet of Things: a comprehensive survey on security challenges, architectures, applications, and future research directions *Trans. Emerg. Telecommun. Technol.* **32** e4337
Li Z, Kang J, Yu R, Ye D, Deng Q and Zhang Y 2017 Consortium blockchain for secure energy trading in industrial internet of things *IEEE Trans. Ind. Inf.* **14** 3690–700
Liu Y, He D, Obaidat M S, Kumar N, Khan M K and Raymond Choo K K 2020 Blockchain-based identity management systems: a review *J. Netw. Comput. Appl.* **166** 102731
Manolache M A, Manolache S and Tapus N 2022 Decision making using the blockchain proof of authority consensus *Proced. Comput. Sci.* **199** 580–8
McGhin T, Choo K K R, Liu C Z and He D 2019 Blockchain in healthcare applications: Research challenges and opportunities *J. Netw. Comput. Appl.* **135** 62–75
Mestiri H, Kahri F, Bedoui M, Bouallegue B and Machhout M 2017 High throughput pipelined hardware implementation of the KECCAK hash function *2016 Int. Symp. on Signal, Image, Video and Communications, ISIVC 2016* pp 282–6
Monrat A A, Schelén O and Andersson K 2019 A survey of blockchain from the perspectives of applications, challenges, and opportunities *IEEE Access* **7** 117134–51
Nabben K and Zargham M 2022 Permissionlessness *Internet Policy Rev.* **11** 1–10
Nakamoto S 2008 Bitcoin: A peer-to-peer electronic cash system. Decentralized Business Review

Neudecker T and Hartenstein H 2018 Network layer aspects of permissionless blockchains *IEEE Commun. Surv. Tutor.* **21** 838–57

Nguyen C T, Hoang D T, Nguyen D N, Niyato D, Nguyen H T and Dutkiewicz E 2019 Proof-of-stake consensus mechanisms for future blockchain networks: fundamentals, applications and opportunities *IEEE Access* **7** 85727–45

Olsen J P, Syed S, Borit M and Boéchat C 2022 Application, limitations, costs and benefits related to the use of blockchain technology in the seafood industry *Report* (Tromsø: Nofima AS) https://nofima.brage.unit.no/nofima-xmlui/handle/11250/2586121

Omote K and Yano M 2020 Bitcoin and blockchain technology *Blockchain and Crypt Currency* 129 (London: SpringerOpen) 129–36

Patel R, Migliavacca M and Oriani M E 2022 Blockchain in banking and finance: a bibliometric review *Res. Int. Bus. Finance* **62** 101718

Peng L, Feng W, Yan Z, Li Y, Zhou X and Shimizu S 2021 Privacy preservation in permissionless blockchain: a survey *Digit. Commun. Netw.* **7** 295–307

Polge J, Robert J and Le Traon Y 2021 Permissioned blockchain frameworks in the industry: a comparison *ICT Express* **7** 229–33

Popovic D, Avis C, Byrne M, Cheung C, Donovan M, Flynn Y, Fothergill C, Hosseinzadeh Z, Lim Z and Shah J 2020 Understanding blockchain for insurance use cases *Br. Actuarial J.* **25** e12

Raikwar M, Mazumdar S, Ruj S, Sen Gupta S, Chattopadhyay A and Lam K Y 2018 A blockchain framework for insurance processes *2018 9th IFIP Int. Conf. on New Technologies, Mobility and Security, NTMS 2018—Proc. (2018-January)* pp 1–4

Saad S M S and Radzi R Z R M 2020 Comparative review of the blockchain consensus algorithm between proof of stake (pos) and delegated proof of stake (dpos) *Int. J. Innov. Comput.* **10** 27–32

Saari A, Vimpari J and Junnila S 2022 Blockchain in real estate: Recent developments and empirical applications *Land Use Policy* **121** 106334

Sankar L S, Sindhu M and Sethumadhavan M 2017 Survey of consensus protocols on blockchain applications *2017 4th Int. Conf. on Advanced Computing and Communication Systems (ICACCS)* pp 1 5

Sanka A I, Irfan M, Huang I and Cheung R C 2021 A survey of breakthrough in blockchain technology: Adoptions, applications, challenges and future research *Comput. Commun.* **169** 179–201

Sarmah S S 2018 Understanding blockchain technology *Comput. Sci. Eng.* **8** 23–9

Schinckus C 2021 Proof-of-work based blockchain technology and anthropocene: an undermined situation? *Renew. Sustain. Energy Rev.* **152** 111682

Sheldon M D 2019 A primer for information technology general control considerations on a private and permissioned blockchain audit *Curr. Issues Audit.* **13** A15–29

Sheth H and Dattani J 2019 Overview of blockchain technology *Asian J. Converg. Technol.* **5** 1–3

Singhal B, Dhameja G, Panda P S, Singhal B, Dhameja G and Panda P S 2018 How blockchain works *Beginning Blockchain: A Beginner's Guide to Building Blockchain Solutions* (Berkeley, CA: Apress) 31–148

Skiba D J 2017 The potential of blockchain in education and health care *Nurs. Educ. Perspect.* **38** 220–1

Sobti R 2012 Cryptographic hash functions: a review *Citeseer* (https://citeseerx.ist.psu.edu/document?repid=rep1&type=pdf&doi=85abc4805adb741b0f8c962794d2ab4dac975c5f)

Solat S, Calvez P and Naït-Abdesselam F 2021 Permissioned vs. permissionless blockchain: how and why there is only one right choice *J. Softw.* **16** 95–106

Sookhak M, Jabbarpour M R, Safa N S and Yu F R 2021 Blockchain and smart contract for access control in healthcare: A survey, issues and challenges, and open issues *J. Netw. Comput. Appl.* **178** 102950

Swati J, Nitin P, Saurabh P, Parikshit D, Gitesh P and Rahul S 2022 Blockchain based trusted secure philanthropy platform: crypto-gocharity *2022 6th Int. Conf. On Computing, Communication, Control and Automation (ICCUBEA)* pp 1–8

Tapscott A and Tapscott D 2017 How blockchain is changing finance *Harv. Bus. Rev.* **1** 2–5

Thompson S 2017 The preservation of digital signatures on the blockchain *See Also* **3** 1–17

Treleaven P, Brown R G and Yang D 2017 Blockchain technology in finance *Computer* **50** 14–7

Vashchuk O and Shuwar R 2018 Pros and cons of consensus algorithm proof of stake. Difference in the network safety in proof of work and proof of stake *Electron. Inform. Technol.* **9** 106–12

Wang Q, Li R and Zhan L 2021 Blockchain technology in the energy sector: From basic research to real world applications *Comp. Sci. Rev.* **39** 100362

Wang Q and Su M 2020 Integrating blockchain technology into the energy sector — from theory of blockchain to research and application of energy blockchain *Comput. Sci. Rev.* **37** 100275

Wang S, Ouyang L, Yuan Y, Ni X, Han X and Wang F Y 2019 Blockchain-enabled smart contracts: architecture, applications, and future trends *IEEE Trans. Syst., Man, Cybernet.: Syst.* **49** 2266–77

Wessling F and Gruhn V 2018 Engineering software architectures of blockchain-oriented applications *2018 IEEE Int. Conf. on Software Architecture Companion (ICSA-C)* pp 45–6

Wu K, Ma Y, Huang G and Liu X 2021 A first look at blockchain-based decentralized applications *Software: Pract. Exp.* **51** 2033–50

Xiao Y, Zhang N, Lou W and Hou Y T 2020 A survey of distributed consensus protocols for blockchain networks *IEEE Commun. Surv. Tutor.* **22** 1432–65

Yang R, Wakefield R, Lyu S, Jayasuriya S, Han F, Yi X, Yang X, Amarasinghe G and Chen S 2020 Public and private blockchain in construction business process and information integration *Autom. Constr.* **118** 103276

Yu M, Sahraei S, Li S, Avestimehr S, Kannan S and Viswanath P 2020 Coded merkle tree: Solving data availability attacks in blockchains *Financial Cryptography and Data Security: 24th Int. Conf., FC 2020 (Kota Kinabalu, Malaysia, February 10–14, 2020)* (Revised Selected Papers) pp 114–34

Zhang A and Lin X 2018 Towards secure and privacy-preserving data sharing in e-health systems via consortium blockchain *J. Med. Syst.* **42** 140

Zhang S and Lee J-H 2020 Analysis of the main consensus protocols of blockchain *ICT Express* **6** 93–7

Zheng G, Gao L, Huang L, Guan J, Zheng G, Gao L, Huang L and Guan J 2021 Decentralized application (dapp) *Ethereum Smart Contract Development in Solidity* (Singapore: Springer) 253–80

Zheng P, Jiang Z, Wu J and Zheng Z 2023 Blockchain-based decentralized application : a survey *IEEE Open J. Comput. Soc.* **4** 121–33

IOP Publishing

Blockchain-based Peer-to-Peer Transactions in Energy Systems

Nnamdi Nwulu and Uyikumhe Damisa

Chapter 2

Real-world blockchains and their applications

Typically, it is not necessary to build a new blockchain for blockchain-based applications. There are numerous real-world blockchains with various characteristics on which new applications can be built. Some of these blockchains have niche areas of application. In this chapter, various real-world blockchain platforms are presented and some areas of application of a few of them are enumerated.

2.1 Introduction

Blockchains have become popular since the emergence of Bitcoin, a widely adopted cryptocurrency. Despite its enormous success, the Bitcoin blockchain suffers from some setbacks. In an attempt to address some of these setbacks, several other blockchains have been developed. These blockchains tend to come with practical features which make them more suitable for certain applications. In the following sections, we discuss the Bitcoin blockchain and other blockchains, and some of their applications.

2.1.1 Bitcoin

The Bitcoin system, which powers Bitcoin cryptocurrency, was created in 2008. Its creator(s) simply describes Bitcoin as a 'Peer-to-Peer electronic cash system' (Nakamoto 2008). Bitcoin is a digital currency that can be sent over the Internet at high speed and at low cost (Zohar 2015). The Bitcoin system is a decentralised peer-to-peer (P2P) network that stores a public ledger containing transactions executed in the system (Vranken 2017). Bitcoin runs on a public blockchain and employs the proof-of-work (PoW) consensus mechanism. High energy consumption (Conti *et al* 2018) is a key downside of the Bitcoin network. Based on the PoW consensus mechanism, multiple miners compete for each block of transactions to be added to the blockchain by expending significant amounts of energy and computing resources. Its native currency is Bitcoin.

2.1.2 Cardano

Cardano is another decentralised blockchain platform, which uses the proof-of-stake (PoS) consensus mechanism. Its design and development are supported by intensive scientific research. Its native currency is the ada cryptocurrency and it is a public blockchain platform. Cardano is working on the development of a smart contract platform that will enable developers to build enterprise-level decentralised applications.

2.1.3 Quorum

The Quorum blockchain is a private network whose roots can be traced to Ethereum. Enterprises can leverage the Ethereum system for the production of blockchain-based applications through the use of Quorum (https://consensys.net/quorum/faq/). Quorum is reported to be a fork of the Ethereum network—it is a permissioned version of the public Ethereum network which offers contract and transaction privacy (https://www.blockchain-council.org/blockchain/what-is-quorum-how-is-it-different-from-other-blockchain/). The blockchain is an enterprise-focused infrastructure which offers features that appeal to commercial entities, particularly financial institutions (https://www.blockchain-council.org/certifications/certified-quorum-expert/). It employs a voting-based consensus mechanism (https://101blockchains.com/quorum-blockchain-use-cases/) and supports smart contracts. Access to transaction history can be restricted on the Quorum blockchain (Quasim *et al* 2020).

2.1.4 EOS

EOS is a public blockchain network that is based on the EOSIO software framework. It employs the delegated PoS consensus algorithm. Organisations and individuals can build and use a broad variety of applications using the public network.

2.1.5 Multichain

Multichain is a fork of Bitcoin Core that has been modified as a permissioned blockchain (https://chainstack.com/protocols/multichain/). It is an enterprise blockchain that enables organisations to quickly develop and deploy decentralised blockchain applications (https://www.multichain.com/)

2.1.6 Tezos

Tezos is an energy-efficient blockchain that enables developers and a variety of other types of users to develop and interact with a network of decentralised applications. It has a modular architecture that enables seamless community-driven upgrades. It supports smart contract development and employs the PoS consensus mechanism.

2.1.7 Stellar

Stellar is a public open-source blockchain that specifically focuses on facilitating payments and asset issuance. It uses the Stellar Consensus protocol, which uses Proof-of-Agreement. The blockchain is energy efficient, cheap, and fast.

2.1.8 Tron

The Tron blockchain is aimed at decentralising the internet.

2.1.9 Ethereum

Despite being a successful and widely accepted application of blockchain technology, Bitcoin is hardly useful beyond its use as electronic cash. Ethereum changed the narrative with the introduction of blockchain-integrated smart contracts. The smart contract facility has made Ethereum and other smart contract-enabled blockchains relevant in many industries, including supply chain, healthcare, education, voting, and insurance, among others. Ethereum offers application developers a platform to build applications that take advantage of the security and decentralisation features of blockchains (Buterin 2016). The Ethereum blockchain employed the PoW consensus mechanism until recently (Kapengut and Mizrach 2023). In 2022, the blockchain moved from PoW to the PoS consensus mechanism, reducing more than 99% of electricity consumption (De Vries 2023). The original Ethereum blockchain is a public blockchain; however, it is possible to have private and consortium nodes that run the Ethereum code (Buterin 2016). Its native currency is Ether. The Ethereum blockchain has led to an increase in the development of decentralised applications. An application that runs on a P2P network of computer nodes rather than the traditional centralised server may be referred to as a decentralised application (DApp; Taş and Tanrıöver 2019). Ethereum has particularly enjoyed patronage with regard to hosting DApps.

2.1.10 Hyperledger

Hyperledger Fabric is a modular, configurable, enterprise-class distributed ledger technology platform that offers flexibility for innovation in the insurance, supply chain, healthcare, and other industries (Lin *et al* 2022).

2.2 Application of real-world blockchains

2.2.1 Applications of Bitcoin

E-voting

Zhao and Chan (2016) designed protocols to enable private electronic voting on the Bitcoin network. The voting problem is such that there are several voters, each of whom wishes to fund one of two candidates. The candidate who receives the most votes is the winning candidate and is guaranteed to receive the funds from the total number of voters. The voters' votes are kept private. Once the final result is revealed, the winner irrevocably receives the funds.

Identity management
Augot *et al* (2017) presented an identity management scheme integrated with the Bitcoin blockchain, which grants users and identity providers shared control over identities.

Credential storage
Ølnes (2016) presents a case for the Bitcoin blockchain to validate persistent documents, analysing a use case of academic certificates stored on the Bitcoin blockchain. The author concludes that vital information can be securely stored in a cost-effective way by storing it on the blockchain.

2.2.2 Applications of ethereum

E-voting
Yavuz *et al* (2018) investigated an e-voting system using smart contract technology based on the Ethereum network. Voters can cast their vote using their Ethereum wallets. Being Ethereum-based, the application benefits from the security features of blockchains. Moreover, blockchains can potentially preserve the voters' privacy and offer transparency to the vote-counting process. Dhulavvagol *et al* (2020) analysed the performance of Parity and Geth, two prominent Ethereum clients. Using a private blockchain network up, the authors investigated a democratic voting application. Their investigation shows that transactions are significantly faster with Parity than with Geth. Decentralised voting has also been explored in Khoury *et al* (2018) on the Ethereum virtual machine using smart contracts. In their system, voters cast their votes using their mobile numbers and are restricted from casting multiple votes. Shukla *et al* (2018) proposed an online voting platform based on the Ethereum blockchain. On their platform, voters are identified by a government-approved number, which the platform ensures that voters can only vote once. Furthermore, the voters' privacy is preserved by ensuring that votes cannot be traced back to their sources. As an online system where voters can securely cast their votes online without being physically present, the turnout of voters is expected to be relatively higher.

Decentralised marketplace
Ranganthan *et al* (2018) used the Truffle development framework to build a decentralised marketplace based on the Ethereum blockchain. The proposed decentralised marketplace addresses the issues of censorship, lack of user privacy, and high platform fees associated with centralised marketplaces.

Sharing economy
Bogner *et al* (2016) proposed a decentralised application that enables the sharing of everyday devices or objects, which is based on the Ethereum network. The application is powered by a smart contract that allows interested users to sign up and rent devices without the involvement of a third-party institution or release of private user data. Renu and Banik (2021) presented a decentralised P2P blockchain-based framework

for ride sharing, which comprises a user-facing interface and a backend based on Ethereum smart contracts. The system recognises three different user roles, viz. the user, the driver, and legal authorities. Jennath *et al* (2019) proposed a blockchain-based scheme that allows individuals to rent out their vacant land as parking lots for a specified duration. In their proposed system, parking lots are represented using non-fungible tokens. According to the authors, a consortium blockchain is best suited for parking leasing; a private Ethereum blockchain is only used for demonstration purposes.

Transportation
Bagloee *et al* (2019) presented a use-case scenario where private road users receive limited mobility credits to encourage a shift to public transportation, thus reducing traffic congestion. They suggest that a blockchain like Ethereum may be useful for enabling mobility credit holders to spend or trade with their credits.

Ride hailing
Kudva *et al* (2020) proposed what they termed the 'Practical Ethereum Blockchain-based Efficient Ride Hailing Service'. The results of their experiments show that less gas is consumed when drivers deploy their own smart contracts. Additionally, passengers pay less while drivers are more profitable, compared to some other existing systems.

Credential validation
BouSaba and Anderson (2019) proposed an automatic degree validation system based on the Ethereum blockchain. Powered by a smart contract at the backend, the front-end interface enables users to view and edit their profile; however, third parties can only view profiles of interest.

Document authenticity and intellectual property
An Ethereum-based, decentralised scheme that immutably timestamps data, certifies its authenticity, and associates it with its owner was proposed by Fallucchi *et al* (2021), which promotes the digitalisation and transparency of government processes.

Tokenisation
In Khan *et al* (2022), a smart contract based on the Ethereum network was used to tokenise sukuk, a financial instrument that offers returns in the same way as bonds. Castellanos *et al* (2017) adopted the Ethereum blockchain for use in the tokenisation and trading of Guarantees of Origin (GoO). A GoO is an electronic document that proves that 1 MWh of electricity was generated from renewable energy sources. Such decentralised systems enable renewable energy consumers to sell their GoO directly to willing buyers without the need for intermediaries.

Healthcare
Chen *et al* (2019) proposed an Ethereum-based medical data management system, which enables patients to exercise rights of ownership over their data while

permitting easy access by hospitals. Multisig is used to ensure that while the patient has control over access to their data, they cannot alter it without authorisation from the hospital.

Tax return

Hossain *et al* (2020) developed a blockchain-based scheme for an automated tax return verification process. The scheme was simulated using Ganachhe, Metamask, and the Remix integrated development environment.

Land registry

Sahni *et al* (2023) presented the idea of a blockchain-based land registry. The proposed framework employs smart contracts at various stages of land registry and is based on the Ethereum blockchain. A blockchain-based land registry offers a tamperproof record of land titles and makes it difficult, if not impossible, to sell a parcel of land to multiple buyers. Biswas *et al* (2021) proposed a blockchain-based platform to facilitate secure land registration in Bangladesh. These systems have the potential to substantially reduce the time taken to sell/buy landed property, eliminate intermediaries, and improve transparency. The authors plan to develop a prototype of the proposed system on the Ethereum blockchain in the future. Aborujilah *et al* (2021) proposed a blockchain-based land registry management framework for Malaysia, which uses Ethereum-based smart contract technology to develop a prototype that achieves the main functionalities of land registration.

Energy

Damisa *et al* (2022) developed an energy auction and reserve trading scheme powered by Ethereum-based smart contracts. In their scheme, buyers and sellers submit their respective offers and bids to a smart contract that performs the auction procedure. Smart contract technology is also employed to facilitate payments to appropriate parties. A similar scheme was developed by Damisa and Nwulu (2022) to facilitate sharing of storage capacity and compensation for shared storage. This scheme can be improved to help shared facility controllers leverage pooled residential storage capacities. Damisa *et al* (2022) proposed an Ethereum smart contract-based scheme to procure gas supplies for a thermal power plant. A smart contract is programmed to receive offers from gas suppliers and bids from a thermal plant operator, after which it can determine the winning offers.

Charity

A blockchain-based scheme based on the Ethereum protocol was proposed by Farooq *et al* (2020) for the transparent, secure, and efficient management of charity funds. Smart contracts are used to enable an exchange of the platform's digital currency, CharityCoin, for fiat currency, the trading of CharityCoin, calls for donations and the transfer of CharityCoin to individuals and organisations. Arantes *et al* (2018) proposed an Ethereum-based scheme for the Brazilian Development Bank, which improves transparency in public funds allocation and supports the monitoring and evaluation of development projects. Ethereum was selected because of its maturity and the ability to express and guarantee the execution of business domain rules.

E-waste management

Gupta and Bedi (2018) proposed an e-waste (discarded electronic and electrical equipment) management system powered by Ethereum-based smart contracts to improve coordination among retailers, producers, importers, and recyclers of electrical and electronic equipment.

2.2.3 Applications of hyperledger

To facilitate international trade, Macedo (2018) proposes using blockchain to create, store, and distribute preferential certificates of origin for preferential trade agreements. With this implementation, the certificate-of-origins registers become immutable. The proposed platform is executed on the Linux Foundation Hyperledger platform.

Tax

Alkhodre *et al* (2019) employed Hyperledger Fabric along with Hyperledger Composer in the design of a value-added tax (VAT) system for the Kingdom of Saudi Arabia. This system provides a tamper-proof and transparent database of VAT transactions, thereby curbing the activities of fraudulent actors. Pasha *et al* (2022) proposed a blockchain-based system, GSTChain, to facilitate an efficient and transparent collection of goods and services tax for the government of India.

Real estate

Gupta *et al* (2019) proposed a land administration system based on the Hyperledger Fabric. The proposed architecture enables land registration, verification, and revocation using specially designed blockchain transactions that are managed by stakeholders. The scheme introduces efficiency, transparency, and security to landed property administration.

Other blockchains

Credential verification

Badr *et al* (2019) developed a system which allows academic institutions to verify and transfer academic records in a secure manner. This eliminates the overhead involved in the traditional manual way of verifying academic records. With this system, academic institutions can efficiently verify and transfer records at a student's request. This system is based on Hyperledger Fabric and uses Hyperledger Composer.

Supply chain

Uddin (2021) proposed a track and trace Hyperledger-based system for the secure and efficient execution of drug supply chain transactions among participating pharmaceutical stakeholders. This integration of such systems with decentralised file systems such as InterPlanetary File System (IPFS) enhances traceability and system transparency.

Healthcare
Due to its enhanced privacy-preserving and data processing capabilities, Wang and Qin (2021) adopted the Hyperledger Fabric system for their proposed healthcare data sharing framework. Similarly, Hyperledger Fabric is used to securely store and transparently share electronic health records among healthcare stakeholders in a P2P network. Uddin *et al* (2021) developed fabric-based chain codes to manage the business logic associated with individual electronic health record transactions on the network. Stamatellis *et al* (2020) achieved a proof-of-concept implementation of a privacy-preserving electronic health record management system using the Hyperledger Fabric blockchain framework. Their proposed scheme stores health records anonymously and can potentially scale for a real-world use case. Antwi *et al* (2021) investigated the use of a private blockchain like Hyperledger Fabric to address privacy and security threats of traditional electronic health record systems. The outcome of their investigation shows the security and scalability of private blockchains in healthcare applications.

E-voting
Kirillov *et al* (2019) presented a Hyperledger Fabric-based e-voting scheme that permits both paper-based voting and e-voting. This scheme offers increased trust and transparency, immutability, and enhanced auditability. Pawlak and Poniszewska-Marańda (2021) implemented an auditable, blockchain-based system on the Hyperledger Fabric network to facilitate a transparent voting process.

Ride hailing
A framework for developing a decentralised ride-hailing architecture was proposed by Shivers *et al* (2021), which is implemented on the Hyperledger Fabric platform. A decentralised ride-hailing platform will enable the participation of autonomous vehicles in ride-hailing because owners can include their vehicles in a community-driven fleet when the vehicle is not being used.

2.3 Summary

The Bitcoin blockchain and its native cryptocurrency Bitcoin has sparked a lot of interest in blockchain technology. It is, however, very limited in its scope of applications, being originally developed as a means of P2P electronic payment system. The Ethereum blockchain improved on Bitcoin by integrating smart contract technology. Smart contract-enabled blockchains find applications in a wide range of areas. The varied characteristics of real-world blockchains make a specific blockchain more suitable for certain applications than others. In addition to carefully assessing the requirements of an intended blockchain application, an assessment of the types of application being built on real-world blockchains can be beneficial in determining the correct blockchain for a project.

References

Aborujilah A, Yatim M N B M and Al-Othmani A 2021 Blockchain-based adoption framework for authentic land registry system in Malaysia *Telkomnika* **19** 2038–49

Alkhodre A, Jan S, Khusro S, Ali T, Alsaawy Y and Yasar M 2019 A blockchain-based value added tax (VAT) system: Saudi Arabia as a use-case *Int. J. Adv. Comput. Sci. Appl* **10** 708–16

Antwi M, Adnane A, Ahmad F, Hussain R, ur Rehman M H and Kerrache C A 2021 The case of HyperLedger Fabric as a blockchain solution for healthcare applications *Blockchain: Res. Appl.* **2** 100012

Arantes G M, D'Almeida J N, Onodera M T, Moreno S M D B M and Almeida V da R S 2018 Improving the process of lending, monitoring and evaluating through Blockchain Technologies: an application of Blockchain in the Brazilian Development Bank (BNDES) *2018 IEEE Int. Conf. on Internet of Things (IThings) and IEEE Green Computing and Communications (GreenCom) and IEEE Cyber, Physical and Social Computing (CPSCom) and IEEE Smart Data (SmartData)* pp 1181–8

Augot D, Chabanne H, Chenevier T, George W and Lambert L 2017 A user-centric system for verified identities on the bitcoin blockchain *Proc. Data Privacy Management, Cryptocurrencies and Blockchain Technology: ESORICS 2017 Int. Workshops, DPM 2017 and CBT 2017 (Oslo, Norway, September 14–15, 2017)* pp 390–407

Badr A, Rafferty L, Mahmoud Q H, Elgazzar K and Hung P C K 2019 A permissioned blockchain-based system for verification of academic records *2019 10th IFIP Int. Conf. on New Technologies, Mobility and Security (NTMS)* pp 1–5

Bagloee S A, Tavana M, Withers G, Patriksson M and Asadi M 2019 Tradable mobility permit with bitcoin and ethereum–a blockchain application in transportation *Internet Things* **8** 100103

Biswas M, Al Faysal J and Ahmed K A 2021 Landchain: a blockchain based secured land registration system *2021 Int. Conf. on Science and Contemporary Technologies (ICSCT)* pp 1–6

Bogner A, Chanson M and Meeuw A 2016 A decentralised sharing app running a smart contract on the ethereum blockchain *Proc. of the 6th Int. Conf. on the Internet of Things* pp 177–8

BouSaba C and Anderson E 2019 Degree validation application using solidity and ethereum blockchain *2019 SoutheastCon* 1–5

Buterin V 2016 Ethereum: platform review *Opportunities and Challenges for Private and Consortium Blockchains* **45** 1–45

Castellanos J A F, Coll-Mayor D and Notholt J A 2017 Cryptocurrency as guarantees of origin: simulating a green certificate market with the ethereum blockchain *2017 IEEE Int. Conf. on Smart Energy Grid Engineering (SEGE)* pp 367–72

Chen H S, Jarrell J T, Carpenter K A, Cohen D S and Huang X 2019 Blockchain in healthcare: a patient-centered model *Biomed. J. Sci. Tech. Res.* **20** 15017

Conti M, Kumar E S, Lal C and Ruj S 2018 A survey on security and privacy issues of bitcoin *IEEE Commun. Surv. Tutor.* **20** 3416–52

Damisa U and Nwulu N I 2022 Blockchain-based auctioning for energy storage sharing in a smart community *Energies* **15** 1954

Damisa U, Nwulu N I and Siano P 2022 Towards blockchain-based energy trading: a smart contract implementation of energy double auction and spinning reserve trading *Energies* **15** 4084

Damisa U, Oluseyi P O and Nwulu N I 2022 Blockchain-based gas auctioning coupled with a novel economic dispatch formulation for gas-deficient thermal plants *Energies* **15** 5155

De Vries A 2023 Cryptocurrencies on the road to sustainability: ethereum paving the way for bitcoin *Patterns* **4** 1–5

Dhulavvagol P M, Bhajantri V H and Totad S G 2020 Blockchain ethereum clients performance analysis considering e-voting application *Procedia Comput. Sci.* **167** 2506–15

Fallucchi F, Gerardi M, Petito M and De Luca E W 2021 Blockchain framework in digital government for the certification of authenticity, timestamping and data property http://hdl.handle.net/10125/70895

Farooq M S, Khan M and Abid A 2020 A framework to make charity collection transparent and auditable using blockchain technology *Comput. Electr. Eng.* **83** 106588

Gupta N and Bedi P 2018 E-waste management using blockchain based smart contracts *2018 Int. Conf. on Advances in Computing, Communications and Informatics (ICACCI)* pp 915–21

Gupta N, Das M L and Nandi S 2019 LandLedger: blockchain-powered land property administration system *2019 IEEE Int. Conf. on Advanced Networks and Telecommunications Systems (ANTS)* pp 1–6

Hossain S, Saha S, Akhi J F and Helaly T 2020 Automated tax return verification with blockchain technology *Proc. of Int. Joint Conf. on Computational Intelligence: IJCCI 2019* pp 45–55

Jennath H S, Adarsh S, Chandran N V, Ananthan R, Sabir A and Asharaf S 2019 Parkchain: a blockchain powered parking solution for smart cities *Front. Blockchain* **2** 6

Kapengut E and Mizrach B 2023 An event study of the ethereum transition to proof-of-stake *Commodities* **2** 96–110

Khan N, Kchouri B, Yatoo N A, Kräussl Z, Patel A and State R 2022 Tokenization of sukuk: Ethereum case study *Glob. Finan. J.* **51** 100539

Khoury D, Kfoury E F, Kassem A and Harb H 2018 Decentralized voting platform based on ethereum blockchain *2018 IEEE Int. Multidisciplinary Conf. on Engineering Technology (IMCET)* pp 1–6

Kirillov D, Korkhov V, Petrunin V, Makarov M, Khamitov I M and Dostov V 2019 Implementation of an e-voting scheme using hyperledger fabric permissioned blockchain *Computational Science and Its Applications–ICCSA 2019: 19th Int. Conf., Saint Petersburg, Russia, July 1–4, 2019, Proc., Part II 19* pp 509–21

Kudva S, Norderhaug R, Badsha S, Sengupta S and Kayes A S M 2020 Pebers: Practical ethereum blockchain based efficient ride hailing service *2020 IEEE Int. Conf. on Informatics, Iot, and Enabling Technologies (ICIOT)* pp 422–8

Lin S-Y, Zhang L, Li J, Ji L and Sun Y 2022 A survey of application research based on blockchain smart contract *Wirel. Netw.* **28** 635–90

Macedo L 2018 Blockchain for trade facilitation: Ethereum, eWTP, COs and regulatory issues *World Customs J.* **12** 87–94

Nakamoto S 2008 Bitcoin: a peer-to-peer electronic cash system *Decentral. Bus. Rev.* 21260

Ølnes S 2016 Beyond bitcoin enabling smart government using blockchain technology *Electronic Government: 15th IFIP WG 8.5 Int. Conf., EGOV 2016, Guimarães, Portugal, September 5–8, 2016, Proc.* 15 pp 253–64

Pasha S H, Mehrotra D, Lin J C-W and Srivastava G 2022 GSTChain: a blockchain network application for the goods and services tax *J. Circuits Syst. Comput.* **31** 2250002

Pawlak M and Poniszewska-Marańda A 2021 Implementation of auditable blockchain voting system with hyperledger fabric *Computational Science–ICCS 2021: 21st Int. Conf., Krakow, Poland, June 16–18, 2021, Proc., Part I* pp 642–55

Quasim M T, Khan M A, Algarni F, Alharthy A and Alshmrani G M M 2020 Blockchain frameworks. decentralised internet of things: a blockchain perspective *Decentralised Internet of Things. Studies in Big Data* Vol 71. (Cham: Springer) 75–89

Ranganthan V P, Dantu R, Paul A, Mears P and Morozov K 2018 A decentralized marketplace application on the ethereum blockchain *2018 IEEE 4th Int. Conf. on Collaboration and Internet Computing (CIC)* pp 90–7

Renu S A and Banik B G 2021 Implementation of a secure ridesharing DApp using smart contracts on Ethereum blockchain *Int. J. Saf. Secur. Eng.* **11** 167–73

Sahni U, Garg S, Srivastava T, Sharma T, Malsa N, Ghosh A, Shaw R N and Gupta V 2023 Framework for land registry system using ethereum blockchain *Advanced Communication and Intelligent Systems: First Int. Conf., ICACIS 2022, Virtual Event, October 20–21, 2022, Revised Selected Papers* pp 431–40

Shivers R, Rahman M A, Faruk M J H, Shahriar H, Cuzzocrea A and Clincy V 2021 Ride-hailing for autonomous vehicles: hyperledger fabric-based secure and decentralize blockchain platform *2021 IEEE Int. Conf. on Big Data (Big Data)* pp 5450–9

Shukla S, Thasmiya A N, Shashank D O and Mamatha H R 2018 Online voting application using ethereum blockchain *2018 Int. Conf. on Advances in Computing, Communications and Informatics (ICACCI)* pp 873–80

Stamatellis C, Papadopoulos P, Pitropakis N, Katsikas S and Buchanan W J 2020 A privacy-preserving healthcare framework using hyperledger fabric *Sensors* **20** 6587

Taş R and Tanrıöver Ö Ö 2019 Building a decentralized application on the ethereum blockchain *2019 3rd Int. Symp. on Multidisciplinary Studies and Innovative Technologies (ISMSIT)* pp 1–4

Uddin M 2021 Blockchain medledger: hyperledger fabric enabled drug traceability system for counterfeit drugs in pharmaceutical industry *Int. J. Pharm.* **597** 120235

Uddin M, Memon M S, Memon I, Ali I, Memon J, Abdelhaq M and Alsaqour R 2021 Hyperledger fabric blockchain: secure and efficient solution for electronic health records *Comp., Mater. Contin.* **68** 2377–97

Vranken H 2017 Sustainability of bitcoin and blockchains *Curr. Opin. Environ. Sustain.* **28** 1–9

Wang Q and Qin S 2021 A hyperledger fabric-based system framework for healthcare data management *Appl. Sci.* **11** 11693

Yavuz E, Koç A K, Çabuk U C and Dalkılıç G 2018 Towards secure e-voting using ethereum blockchain *2018 6th Int. Symp. on Digital Forensic and Security (ISDFS)* pp 1–7

Zhao Z and Chan T-H H 2016 How to vote privately using bitcoin *Information and Communications Security: 17th Int. Conf., ICICS 2015 (Beijing, China, December 9–11, 2015) Revised Selected Papers 17* pp 82–96

Zohar A 2015 Bitcoin: under the hood *Commun. ACM* **58** 104–13

Nnamdi Nwulu and Uyikumhe Damisa

Chapter 3

A literature review of recent advances in blockchain technology

Blockchain technology lies at the intersection of innovation and challenges. Renowned for decentralising digital interactions, blockchain's applications extend from underpinning cryptocurrencies to revolutionary financial models such as Decentralised Finance (DeFi). It offers unprecedented traceability in supply chains, individual empowerment through digital identities, protection of democratic processes, and advances in healthcare. However, the path to its widespread acceptance is dotted with hurdles, including scalability, energy concerns, security vulnerabilities, and societal and legal intricacies. Reflecting on these dimensions, the trajectory of blockchain appears to be promising yet complex. Future research should address these challenges and dive into sector-specific applications, technological advancements, and global regulatory landscapes. Blockchain's transformative potential beckons to a future where transparency, security, and decentralisation are pillars of our digital ecosystem.

3.1 Introduction

Blockchain technology, has rapidly transformed from its obscure origins to a global phenomenon. It was initially designed as a distributed ledger or database shared across a vast network of computers. Every piece of data within it is grouped into a block and, once verified, links to the previous block, forming an interconnected chain that is aptly named 'blockchain.' Its roots can be traced to cryptographic advances such as public-key cryptography in the late-1970s and 1980s. But it was not until 2008 that the term 'blockchain' gained prominence, introduced by an entity or perhaps an individual named Satoshi Nakamoto through the Bitcoin white paper (Nakamoto 2008). Bitcoin, the first application of blockchain technology, showcased the potential

doi:10.1088/978-0-7503-6295-5ch3

of a decentralised ledger to facilitate direct transactions without intermediaries. The subsequent years saw an evolution from Bitcoin's singular use to many applications, most notably Ethereum in 2015, which integrated 'smart contracts' into its platform (Abdulhakeem and Hu 2021). The potential applications eventually began to extend far beyond currency, reaching a wide range of industries.

At the heart of blockchain lies its structure and cryptographic principles. Each block contains a set of transactions, a timestamp, and a cryptographic hash of its preceding block. This architecture ensures that all data remains interconnected. Two primary cryptographic methods support this structure. The first is the hash function, which converts input data into a fixed-length string, so any slight change in input dramatically transforms the output. The second is public-key cryptography, wherein users generate a pair of keys: one public and shared, and another private and kept secret. While the public key encrypts the data, only its corresponding private key can decrypt it. Several principles such as decentralisation, immutability, and consensus algorithms are fundamental to the blockchain. Unlike traditional centralised systems, the blockchain operates on numerous nodes, each possessing a copy of the entire ledger. Once a block integrates into the chain, altering it becomes almost impossible (Wang *et al* 2021). To ensure synchronisation across all nodes, consensus mechanisms such as the Proof-of-Work (PoW) come into play.

Think of a block as a ledger page, the blockchain is the entire book. A continuous chain forms, with each block filled with data and pointing to its predecessor through a cryptographic hash. This decentralised approach is one of the core strengths. Transparency and trust are inherent, with no central authority and every participant having access to the entire database and its full history. Consider the analogy of sending a sealed envelope for a more vivid depiction of its cryptographic principles. Only the unique keyholder can open this envelope, much like encrypted data with a public key being decrypted only by its corresponding private key. Initially, the main use of blockchain was in Bitcoin, a groundbreaking digital currency. However, as the technology matured, its applications diversified into finance, supply chains, and voting systems, highlighting its versatility. In today's context, the importance of blockchain cannot be understated. The past decade has seen it evolve with the advent of decentralised finance, non-fungible tokens (NFTs) and other innovations. Although early networks grappled with scalability, new solutions emerged to address these challenges. Transaction processing times have improved, security remains a cornerstone given the technology's decentralised nature, and developments such as Zero-Knowledge Succinct Non-Interactive Argument of Knowledge (zk-SNARKs) have enhanced transactional privacy without compromising security. In this digital age, which is marked by increasing concerns about data privacy, trust, and security, blockchain has emerged as a significant player (Javaid *et al* 2021). This review aims to help the reader to delve into blockchain's nuances, from its historical context to its current challenges and prospects, offering a holistic understanding of this revolutionary technology.

3.2 Methodology

For a comprehensive bibliometric study of blockchain technology, the Scopus database was visualised. These platforms were selected due to their extensive coverage across various disciplines, offering a wide spectrum of available literature. A mixture of keywords and search strings was used to maximise search efficacy. Keywords such as 'blockchain,' 'cryptocurrency,' 'decentralised ledger', 'smart contracts,' 'blockchain applications,' and 'blockchain challenges' proved instrumental. These keywords were combined using Boolean operators such as AND, OR, and NOT, creating search strings such as 'blockchain AND decentralised ledger' and 'blockchain NOT cryptocurrency.'

The literature search strategy initiated foundational searches using individual keywords to yield broad results. From this foundational stage, keywords were combined to refine the scope and identify relevant articles. For instance, literature highlighting various blockchain applications was targeted by coupling 'blockchain' with 'applications.' Specific inclusion and exclusion criteria were established to maintain the relevance and quality. The focus was on articles published within the last 10 years to ensure current information. Furthermore, only peer-reviewed articles were considered, emphasising the credibility of the sourced literature. The emphasis was placed on articles that explored blockchain's applications, challenges, and potentials (Shoaib *et al* 2022). Conversely, non-peer-reviewed articles, literature published more than a decade ago, or pieces not written in English were excluded due to concerns of academic rigour, relevance, and potential translation challenges. Figures 3.1–3.3 show a distribution of publications by country, co-occurrence of keywords, and publications per author, respectively.

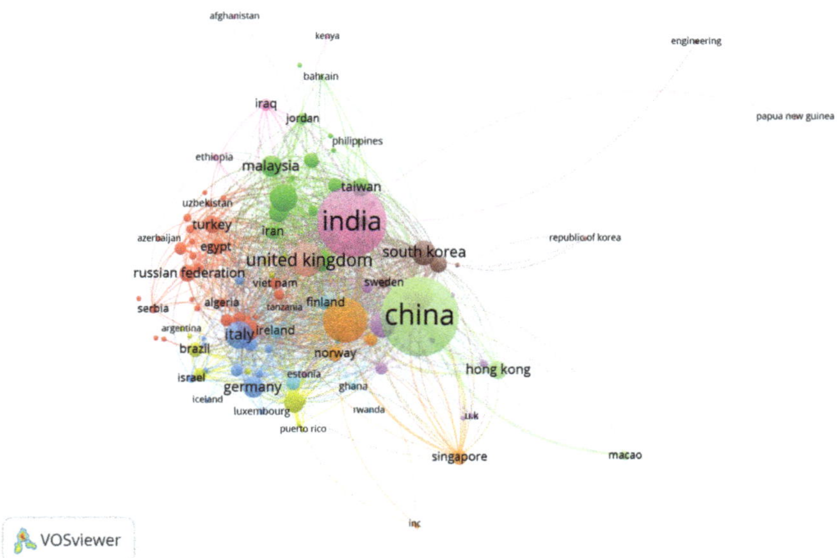

Figure 3.1. Publications per country.

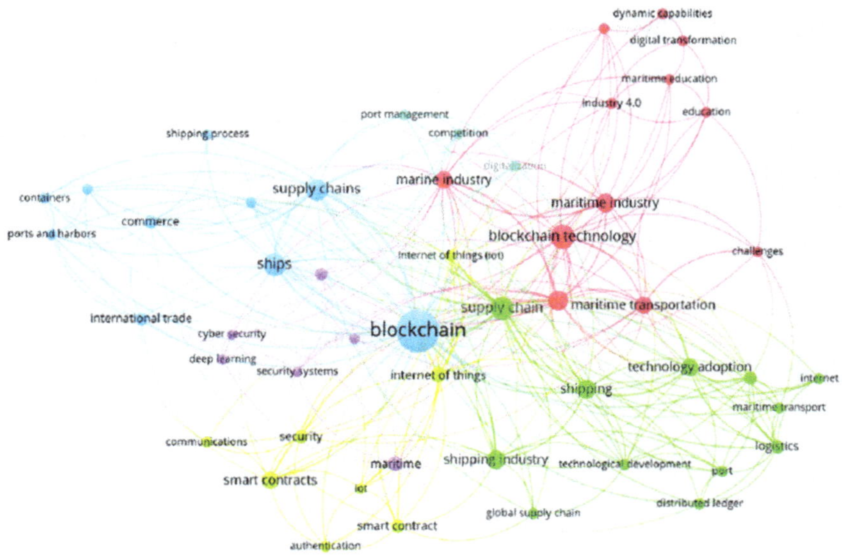

Figure 3.2. Keywords of co-occurrence.

Figure 3.3. Publications per author.

After compiling a list of potential articles, the titles and abstracts underwent screening for relevance. This was followed by a full text review of shortlisted articles, determining their inclusion or removal based on established criteria. An objective two-reviewer system ensured that two different reviewers independently assessed each article. In disagreements between the reviewers, a consensus was sought, or a third reviewer was consulted. Once the selection process was completed, the literature was organised on the following themes: foundational principles of blockchain, its varied applications, the challenges encountered, and potential forward-looking solutions. This thematic approach facilitated systematic exploration and analysis, presenting a detailed overview of blockchain's various aspects (Niknejad *et al* 2020). This methodology aims to provide a comprehensive perspective on the current state of blockchain research and its potential directions.

3.3 Discussion: review of recent advances in blockchain technology

In recent years, the technological world has marvelled at blockchain technology's rapid transformation and growth. At its heart, the blockchain represents a ledger system that allows digital information to be distributed without being copied or altered illegitimately. This technology has shown a myriad of enhancements in both its capabilities and the scope of its applications, prompting interest from sectors as diverse as finance to healthcare. A central component to understanding the evolution of blockchain technology lies in its consensus mechanisms. At a basic level, a consensus mechanism refers to the technique by which a blockchain network agrees on the ledger's contents. This ensures that every transaction recorded on the blockchain is verified and agreed upon by all participants in the network. Historically, the most well-known consensus mechanism has been the PoW system. Bitcoin, the pioneer of cryptocurrencies, operates on this PoW mechanism. In PoW, participants (or miners) solve complex mathematical problems to validate and add transactions to the blockchain. However, this method has drawn significant criticism due to its enormous energy consumption. Some studies even equate the energy consumption of the Bitcoin network with the energy needs of entire countries, underscoring the environmental concerns related to using PoW as a consensus mechanism (Xiong *et al* 2022).

Such environmental concerns have sparked the search for alternative consensus mechanisms that are more energy efficient. One of the emerging favourites in this space is the Proof-of-Stake (PoS) mechanism. In stark contrast to the computational challenges posed by PoW, PoS decides on the individual or group who gets to validate the next block of transactions based on the number of coins they hold and are willing to 'stake' or lock up as collateral. Ethereum, which is one of the premier platforms in the cryptocurrency world, has announced plans to transition from PoW to PoS, reinforcing the appeal and potential of PoS (Xiong *et al* 2022). However, the world of consensus mechanisms does not end with PoS. An offshoot of PoS, termed Delegated Proof of Stake (DPoS), is steadily gaining traction. In the DPoS system, rather than every coin holder participating in the validation process, they elect a few representatives or delegates. These delegates are responsible for validating transactions and maintaining the blockchain's integrity. EOS is a prime example of a platform that champions the DPoS system. The rationale behind DPoS is that by limiting the number of participants in the consensus process, transactions can be validated more quickly, improving the blockchain's scalability and efficiency (Xiong *et al* 2022).

Moving further into the realm of consensus mechanisms, Byzantine Fault Tolerance (BFT)-based mechanisms, such as the Federated Byzantine Agreement (FBA) and Practical Byzantine Fault Tolerance (PBFT) have surfaced as innovative solutions. Stellar's usage of FBA and Hyperledger Fabric's utilisation of PBFT are emblematic of this emerging trend, highlighting the diversity of approaches to achieve consensus in a decentralised network (Scott *et al* 2023). However, as the blockchain universe expands and diversifies, it faces the perennial challenge of scalability. In essence, as more users join the network and the volume of transactions increases, the speed and efficiency of the network can be hampered. Layer2 solutions, which are protocols built on top of the primary blockchain have been developed to enhance scalability. By conducting most transactional processes off the main blockchain, these solutions promise to alleviate network congestion and reducing transaction fees. Bitcoin's Lightning Network and Ethereum's Plasma are salient examples in this category. Additionally, the concept of'sharding', wherein the blockchain is segmented into smaller, manageable parts called'shards', is another ingenious solution. Ethereum 2.0, with its ambitious roadmap, intends to employ sharding to enhance its scalability (Scott *et al* 2023).

A natural progression from the scalability discussion is the topic of interoperability. As numerous blockchains with diverse applications emerge, the ability of these blockchains to communicate and interact with each other becomes crucial. Projects such as Polkadot and Cosmos, who are pioneers in this space, have embarked on missions to connect different blockchains seamlessly. Polkadot's approach involves a relay chain coordinating multiple para-chains, while Cosmos opts for a hub-and-spoke model to facilitate inter-blockchain communication (Scott *et al* 2023). Parallel to these technological developments is the increasing emphasis on sustainability. The criticism aimed at PoW's energy-intensive operations has galvanised efforts to transition to more eco-friendly consensus mechanisms. Ethereum's pivot to PoS exemplifies this shift, marking a clear move towards prioritising environmental sustainability in the blockchain space. DPoS and BFT-based mechanisms, with their more optimised and efficient methodologies, further champion this cause (Xiong *et al* 2022).

As blockchain networks become more complex, their vulnerability to security breaches increases. This has spurred innovations in cryptographic techniques, with zero-knowledge proofs (ZKPs) leading the way. One variant, zkSNARKs, employed by Zcash, strikes a fine balance by offering enhanced transactional privacy without compromising on network security, illustrating the groundbreaking advances in blockchain security measures (Teng *et al* 2023). On the application side, the blossoming of DeFi platforms is reshaping the financial industry. Using blockchain technology, platforms like Uniswap, Aave, and Compound democratise financial systems, reducing the dependency on traditional banking intermediaries. The rise of non-fungible tokens (NFTs) is another fascinating development, revolutionising the way in which we perceive digital asset ownership. Meanwhile, Central Bank Digital Currencies (CBDCs), such as China's DCEP, showcase the immense potential of integrating blockchain technology into national financial infrastructures (Teng *et al* 2023).

However, even as the blockchain space evolves, threats loom. The impending age of quantum computing threatens to make existing cryptographic techniques obsolete. Recognising this, initiatives such as the Quantum Resistant Ledger (QRL) are pioneering efforts to build blockchains resilient to quantum threats (Teng *et al* 2023). Complementing the blockchain's growth story is its convergence with other emerging technologies. Projects such as SingularityNET, which marries AI with blockchain, or IOTA's Tangle technology, which is tailored for the Internet of Things (IoT), exemplify the limitless possibilities when technologies intersect (Wu *et al* 2022). In summary, the ever-evolving landscape of blockchain technology, with its continuous innovations, serves as a testament to its adaptability and growing relevance in the contemporary digital world. The dynamism inherent in its growth narrative reflects the potential of blockchain to redefine the digital frontier in the years to come (Wu *et al* 2022).

3.4 The evolving applications of blockchain technology

Blockchain technology, at its core, is a decentralised, distributed, and often public digital ledger that is made up of records known as blocks. It is underpinned by key principles such as decentralisation, transparency, and security. The implications of these principles have reverberated across multiple industries and brought about transformative changes. The primary and well-recognised application of blockchain is in the realm of cryptocurrencies. It all began with Bitcoin, which was revolutionary not just as a decentralised currency but also in its use of blockchain to verify and record transactions (Lei *et al* 2021). Following Bitcoin's introduction, many other cryptocurrencies have emerged, each serving functions such as mediums of exchange, stores of value, and units of account.

DeFi is an offshoot that represents a seismic shift in how we view financial transactions. By leveraging blockchain's inherent traits, DeFi endeavours to dismantle traditional financial intermediaries. The landscape of DeFi is vast, encompassing stablecoins, decentralised exchanges, and innovative platforms for lending, borrowing, and insurance. Smart contracts are crucial to many DeFi platforms, which are self-executing contracts with the terms of the agreement directly written into code lines. In the creative sphere, NFTs have made substantial inroads. While NFTs have gained notable traction within the art community, their applications stretch to music, real estate, and many virtual goods (Bhutta *et al* 2021). The notion of digital scarcity is central to NFTs, which ensures the uniqueness and verifiability of digital assets.

Supply chain management is yet another area where blockchain has shown immense promise. With its ability to offer end-to-end traceability and transparency, the blockchain is invaluable in areas such as food safety, ensuring fair trade practices, and maintaining the integrity of pharmaceutical products. Another pressing modern concern is digital identity verification. The blockchain offers a platform for secure, immutable, and self-sovereign identities. Such a system not only places individuals in control of their data but also presents a formidable defence against identity theft. The democratic process will gain immensely from blockchain, with the technology offering the potential for secure, transparent, and untameable voting systems (Gad *et al* 2022). However, the journey has not been without its hurdles, including concerns about ensuring voter privacy and accessibility to all eligible voters.

The IoT world is vast and rapidly expanding. By integrating blockchain, there is potential for heightened data security, increased device autonomy, and streamlined device-to-device interactions. CBDCs have recently garnered significant interest. With central banks mulling over the idea of launching their digital renditions of national currencies using blockchain, there is potential for enhanced financial inclusivity, streamlined cross-border transactions, and a more nuanced approach to monetary policy. Finally, blockchain's potential is palpable in the realm of healthcare. From securely managing patient records to ensuring drug traceability and even bolstering medical research, blockchain can revolutionise healthcare systems (Habib *et al* 2022). Blockchain's versatility and robustness make it a cornerstone technology with potential applications in virtually every modern industry. The areas discussed represent the tip of the iceberg in blockchain's transformative journey.

3.5 Challenges of blockchain technology

3.5.1 Technical challenges

While revolutionary, blockchain technology comes with challenges that have impeded its mainstream adoption and scalability. One of the primary issues plaguing blockchains is scalability. The current throughput of many blockchains in terms of transactions per second needs to catch up to traditional payment systems such as Visa and Mastercard. This limited throughput becomes even more evident during peak transaction times, leading to network congestion. This congestion slows the transaction speed and can increase transaction costs, making it less appealing for everyday use. Moreover, there is an inherent trade-off between block size and time. Bigger blocks can hold more transactions, but take longer to propagate through the network (Li *et al* 2021). This extended propagation time can increase the risk of forking the blockchain, creating multiple versions of the transaction history.

A notable concern, especially in environmental sustainability, is the energy efficiency of blockchains, particularly those that rely on the PoW consensus mechanism, such as Bitcoin. These blockchains consume vast amounts of energy, sometimes comparable to the energy consumption of entire nations, thereby exerting a significant environmental toll. While security is one of blockchain's key selling points, it also presents challenges. The possibility of a 51% attack, where a single entity gains majority control over the network's hashing power, poses a substantial risk. This dominance can enable the entity to double-spend or manipulate transactions. Furthermore, while smart contracts have the potential to automate and streamline processes, they are not impervious to bugs and vulnerabilities (Yaqoob *et al* 2021). Notable incidents such as the DAO attack on Ethereum underscore the risks associated with smart contract vulnerabilities.

Though lauded for its ability to ensure accountability, the transparent nature of public blockchains can be a double-edged sword. This transparency can sometimes infringe on user privacy. Efforts to introduce private transactions on these public blockchains come with challenges, although advances such as zkSNARKs and zkSTARKs have made significant strides in preserving privacy. Interoperability, or the ability of different blockchains to communicate and share information seamlessly, remains an area for improvement. The lack of standardisation across blockchain platforms exacerbates this challenge. Usability issues also hinder the broader adoption of blockchain technology. Dealing with cryptographic keys, grasping the intricacies of transaction fees, and navigating the often complex landscape of decentralised applications (DApps) can be daunting for the average user. The stability of blockchain networks is not a given. Rapid growth, contentious forks, or significant changes in the underlying protocol can induce network instability, undermining user trust. Data storage and management presents another challenge. As the volume of transactions on blockchains grows, storing and managing these data efficiently become increasingly complex and resource-intensive. Finally, while still on the horizon, the emergence of practical quantum computing poses a potential threat to blockchains. Quantum computers can break the cryptographic algorithms underpinning blockchains' security, rendering them vulnerable (Ali *et al* 2021). In conclusion, while blockchain offers transformative potential across industries, these challenges underscore the need for continued research, development, and collaboration to realise its full promise.

3.5.2 Examination of societal and legal challenges

Blockchain, which is often regarded as the foundational technology of cryptocurrencies, is a decentralised digital ledger that records transactions across numerous computers. Beyond its application in cryptocurrencies, the unique features of blockchain, such as its transparency, security, and immutability, have been recognised in a wide range of sectors, ranging from healthcare to real estate. However, as its potential applications continue to expand, various societal and legal challenges accompany its adoption. The current regulatory landscape is a significant barrier to more extensive blockchain integration. Given the technology's relative novelty, many countries still need to establish clear guidelines, leading to widespread confusion and uncertainty. The definition of cryptocurrencies is a point of contention: Are they currencies, securities, commodities, or a unique category in their own right? This ambiguity extends to other legal areas, including securities law, tax law, and broader compliance regulations. Moreover, there is a notable global regulation disparity (Weaver 2020). While places such as Switzerland present a conducive environment for blockchain, others, such as China, have implemented outright bans, creating challenges for global interoperability.

Simultaneously, the intersection of blockchain's immutability and regulations, such as the EU's General Data Protection Regulation (GDPR), which mandates the 'right to be forgotten,' presents challenges. While blockchain transparency is often considered to be an asset, its pseudonymous nature is susceptible to exploitation. This potential for misuse in activities such as money laundering or financing terrorism has raised genuine societal and legal concerns. The legal recognition and enforcement of blockchain-based transactions and smart contracts is another area of ambiguity. Some regions are progressive, deeming them legally binding, but many remain undecided (Bury *et al* 2020). A key question is whether these self-executing contracts, which have their terms encoded directly into lines of code, can be challenged or refuted in conventional legal systems.

The resistance of society to blockchain, fuelled partly by misunderstandings or a lack of knowledge, is also a significant challenge. Negative publicity stemming from events such as cryptocurrency thefts and dubious Initial Coin Offerings (ICOs) only reinforces these misconceptions, acting as a roadblock to widespread adoption. The digital divide casts a shadow over blockchain's decentralisation promises. Unequal access to essential infrastructure, such as stable internet and computing resources, and a disparity in technical literacy raise concerns over fair access to blockchain's benefits. There are also ethical dilemmas to consider. The idea of a completely decentralised system prompts questions about accountability and responsibility, especially in the face of potential job displacement due to automation. Finally, the Intellectual Property (IP) realm is not immune to blockchain's challenges. The rise of NFTs brings IP enforcement and protection issues to the forefront. Although NFTs can authenticate digital ownership, they also introduce potential challenges related to IP infringement (Pennycook and Rand 2021). In conclusion, while blockchain undeniably offers revolutionary potential across multiple sectors, it is equally apparent that considerable societal and legal challenges will mark its journey. Navigating these hurdles will require collective efforts from regulators, technologists, and the broader society to harness the full benefits of blockchain while ensuring the protection of individual rights and societal norms.

3.6 Potential future developments in blockchain technology based on current trends

Since its inception, blockchain technology has experienced significant developments. On the basis of current trends, we can project several potential future developments in this field that will shape the digital landscape in the coming years. One of the most pressing concerns surrounding blockchain is scalability and efficiency. As the number of transactions and participants grows, so does the strain on the network. Ethereum, one of the most significant blockchain platforms, is shifting from its current PoW consensus mechanism to PoS with Ethereum 2.0. This transition promises greater scalability and energy efficiency. Layer-2 scaling solutions, such as Optimism and zkSync, are also being developed to offload transaction processing from the mainchain, thereby improving throughput (Levis *et al* 2021). Sharding, a method that divides a blockchain's database into smaller pieces, is also emerging as a promising solution for distributing workload and enhancing scalability.

Interoperability has become a pivotal area of focus. As the number of blockchain networks proliferates, the ability of these networks to communicate and interact becomes vital. Initiatives such as Polkadot and Cosmos aim to break down these silos, allowing for seamless cross-chain communication. Such developments point towards a more interconnected and interoperable blockchain ecosystem. The complexity of smart contracts supporting these platforms is also increasing with the increasing popularity of DeFi and NFT. This necessitates the evolution of smarter and more secure contracts and rigorous testing frameworks (Schlecht *et al* 2021). As for decentralised applications, the focus will likely shift towards developing more user-friendly interfaces and seamless integrations, making the decentralised world more accessible to mainstream users.

On the regulatory front, the crypto-space is still seen as the 'Wild West' in many respects. However, as blockchain becomes a staple in global economies, it is plausible that we will see more structured regulatory guidelines established across jurisdictions, providing clarity and security for developers and users. The institutional gaze is now firmly fixed on the blockchain. Large businesses, banks, and even governments are keenly exploring blockchain's potential, hinting at a future where these entities might incorporate blockchain solutions for various purposes, from supply chain management to financial systems. The concept of CBDCs is no longer relegated to speculative discussions (Merlo *et al* 2022). Many nations are already in the pilot phase and CBDCs could co-exist alongside traditional currencies in the future, offering a digital and decentralised counterpart.

Privacy is paramount in the digital age. Anticipated advances in cryptographic techniques, such as zero-knowledge proofs and homomorphic encryption, could redefine the privacy standards on public blockchains, allowing transparent and confidential transactions. Another crucial development horizon is the quantum computing frontier. As these supercomputers become more viable, they pose risks to the current cryptographic safeguards of the blockchain. Thus, a shift towards developing quantum-resistant blockchains might become indispensable. Finally, sustainability is currently the most pressing issue. With environmental concerns being raised, especially around energy-intensive PoW mechanisms, the future will likely witness a drive towards more eco-friendly consensus mechanisms and practises. In conclusion, blockchain's future seems robust, diversified, and integrated (Alfa *et al* 2020). As it continues to evolve, intersecting with various technologies and sectors, it promises to reshape our understanding of digital interactions, finance, and decentralisation.

3.7 Summary

Blockchain technology, renowned for its pioneering approach to decentralised record-keeping, has presented a new lens through which we view digital interactions and the broader digital realm. In this review, we unearthed a wealth of insights. We delved deep into the essence of blockchain, exploring its fundamental principles like decentralisation, transparency, and security, and how they have been a game-changer across various industries. From being the backbone of cryptocurrencies to catalysing revolutionary financial models such as DeFi, blockchain's multifaceted applications are evident. Blockchain promises enhanced traceability in supply chains, the empowerment of individuals through digital identities, the protection of democratic processes such as voting, and significant strides in healthcare. However, these promises are accompanied by formidable challenges. The struggles with scalability, concerns about energy consumption, potential security breaches, and the delicate balance between privacy and transparency are pertinent issues that underline the technology's evolving nature. On top of these, the societal and legal conundrums, spanning from regulatory grey areas and conflicts with data protection norms to the complexities surrounding intellectual property in the age of NFTs, emphasise that the journey of blockchain's widespread acceptance is multifaceted.

Reflecting on the research questions introduced at the onset, it becomes clear that blockchain, while brimming with potential, remains a dynamic technology. Its trajectory, adoption rate, and seamless integration into daily life depend on the resolution of intrinsic challenges. Looking forward, there are several intriguing avenues. A more granular exploration of individual sectors could shed light on the specific challenges and solutions that blockchain might offer in each domain. Technological advances are eagerly awaited, especially those that address scalability and energy concerns. Given the disparate global attitudes towards blockchain and its manifestations, such as cryptocurrencies, an in-depth study of the regulatory tapestry worldwide would be invaluable. Moreover, as the blockchain further embeds itself in our systems, understanding its societal repercussions, from the job market to economic structures, becomes imperative. In conclusion, the allure of blockchain as a trailblazing technological marvel is undeniable. Although it promises a future where concepts such as transparency, security, and decentralisation are commonplace, its intricate journey ahead calls for persistent research, innovation, and global discourse to truly harness its transformative potential.

References

Abdulhakeem S A and Hu Q 2021 Powered by Blockchain Technology, DeFi (decentralized finance) strives to increase financial inclusion of the unbanked by reshaping the world financial system *Modern Econ.* **12** 1–16

Alfa A A, Alhassan J K, Olaniyi O M and Olalere M 2020 Blockchain technology in IoT systems: current trends, methodology, problems, applications, and future directions *J. Reliab. Intell. Environ.*

Ali O, Jaradat A, Kulakli A and Abuhalimeh A 2021 A comparative study: blockchain technology utilization benefits, challenges and functionalities *IEEE Access* **9** 12730–49

Bhutta M N M, Khwaja A A, Nadeem A, Ahmad H F, Khan M K, Hanif M, Song H, Alshamari M and Cao Y 2021 A survey on blockchain technology: evolution, architecture and security *IEEE Access* **9** 1

Bury S M, Flower R L, Zulla R, Nicholas D B and Hedley D 2020 Workplace social challenges experienced by employees on the autism spectrum: an international exploratory study examining employee and supervisor perspectives *J. Autism Dev. Disord.* **51** 1614–27

Gad A G, Mosa D T, Abualigah L and Abohany A A 2022 Emerging trends in blockchain technology and applications: a review and outlook *J. King Saud Univ.—Comput. Inform. Sci.* **34** 6719–42

Habib G, Sharma S, Ibrahim S, Ahmad I, Qureshi S and Ishfaq M 2022 Blockchain technology: benefits, challenges, applications, and integration of blockchain technology with cloud computing *Future Internet* **14** 341

Javaid M, Haleem A, Pratap Singh R, Khan S and Suman R 2021 Blockchain technology applications for Industry 4.0: a literature-based review *Blockchain: Res. Appl.* **2** 100027

Lei N, Masanet E and Koomey J 2021 Best practices for analyzing the direct energy use of blockchain technology systems: review and policy recommendations *Energy Policy* **156** 112422

Levis D, Fontana F and Ughetto E 2021 A look into the future of blockchain technology *PLOS One* **16** e0258995

Li X, Zheng Z and Dai H-N 2021 When services computing meets blockchain: challenges and opportunities *J. Parallel Distrib. Comput.* **150** 1–14

Merlo V, Pio G, Giusto F and Bilancia M 2022 On the exploitation of the blockchain technology in the healthcare sector: a systematic review **213** *Expert Syst. Appl.* 118897

Nakamoto S 2008 Bitcoin: A peer-to-peer electronic cash system. Decentralized Business Review

Niknejad N, Ismail W, Bahari M, Hendradi R and Salleh A Z 2020 Mapping the research trends on blockchain technology in food and agriculture industry: a bibliometric analysis *Environ. Technol. Innov.* **21** 101272

Pennycook G and Rand D G 2021 Research note: examining false beliefs about voter fraud in the wake of the 2020 Presidential Election *Harvard Kennedy School Misinform. Rev.* **2**

Schlecht L, Schneider S and Buchwald A 2021 The prospective value creation potential of Blockchain in business models: a Delphi study *Technol. Forecast. Soc. Change* **166** 120601

Scott I J, de Castro Neto M and Pinheiro F L 2023 Bringing trust and transparency to the opaque world of waste management with blockchain: a Polkadot parathread application *Comput. Ind. Eng.* **182** 109347

Shoaib M, Zhang S and Ali H 2022 A bibliometric study on blockchain-based supply chain: a theme analysis, adopted methodologies, and future research agenda *Environ. Sci. Pollut. Res.* **30** 14029–49

Teng Y, Luo F, Wu Q and Ranzi G 2023 Blockchain in smart grids: a review of recent developments **44** 23–59

Wang N, Xu H, Xu F and Cheng L 2021 The algorithmic composition for music copyright protection under deep learning and blockchain *Appl. Soft Comput.* **112** 107763

Weaver R L 2020 The impact of COVID-19 on the social enterprise sector *J. Soc. Entrepreneurship* **14** 177–85

Wu C, Xiong J, Xiong H, Zhao Y and Yi W 2022 A review on recent progress of smart contract in blockchain *IEEE Access* **10** 50839–63

Xiong H, Chen M, Wu C, Zhao Y and Yi W 2022 Research on progress of blockchain consensus algorithm: a review on recent progress of blockchain consensus algorithms *Future Internet* **14** 47

Yaqoob I, Salah K, Jayaraman R and Al-Hammadi Y 2021 Blockchain for healthcare data management: opportunities, challenges, and future recommendations *Neural Comput. Appl.* **34** 1–16

Chapter 4

Blockchain applications by industry

Blockchain technology has emerged as a groundbreaking innovation with implications spanning various industries, from healthcare and education to media and government services. Its unique attributes, such as transparency, immutability, and decentralisation, offer solutions tailored to address industry-specific challenges, promoting higher efficiency, trustworthiness, and direct user engagement. While the potential benefits of blockchain integration are immense, the journey towards its widespread adoption is laden with challenges such as scalability issues, regulatory concerns, and the requirement for a deeper public understanding. This chapter provides a comprehensive discussion on the potential benefits and drawbacks of blockchain adoption across different sectors, emphasising the need for a holistic approach in leveraging the technology. The eventual success of blockchain integration will hinge on collaborative efforts among industry stakeholders, regulatory bodies, and the wider public, ultimately shaping the blueprint for our technological future.

4.1 Introduction

Blockchain technology, often hailed as an ingenious invention, has charted a fascinating course from its origins to its modern applications. Tracing its roots back to 1991, Stuart Haber and W. Scott Stornetta envisioned a cryptographically secure chain of blocks, laying the groundwork for ensuring unalterable timestamps on documents. This groundbreaking idea saw a crucial improvement in 1992 when Merkle Trees were integrated, allowing efficient grouping of multiple document certificates in a single block. However, despite its potential, the concept remained largely dormant as the digital world rapidly expanded. It was only in 2008 that blockchain received the attention that it was due (Sunny *et al* 2022). An enigmatic individual or group known as Satoshi Nakamoto unveiled Bitcoin as a decentralised currency system that leaned heavily on blockchain for its transparent and verifiable transactions, eliminating the need for a central authority.

With the birth of Bitcoin in 2009, the digital realm witnessed the actualisation of blockchain, as Nakamoto mined the inaugural block and set up a reward system for miners. The success of Bitcoin paved the way for a deluge of digital currencies, such as Litecoin and Ripple. Although each introduced unique features, the foundational architecture of the blockchain remained consistent. The technological landscape witnessed another shift with the emergence of Ethereum, which not only introduced a new cryptocurrency but also heralded the concept of smart contracts and decentralised applications (DApps), pushing the boundaries of blockchain applications (Renduchintala *et al* 2022). Following Ethereum's debut, the technology found diverse applications, ranging from finance and real estate to supply chain management, and beyond.

At its essence, blockchain is a decentralised digital ledger that records transactions across multiple computers, safeguarding the data's permanence and inalterability. Picture a traditional ledger, now distributed across a vast network, ensuring transparency instead of being confined to a singular entity like a bank. When a transaction is initiated, it is broadcast to a plethora of nodes for validation. Once validated, it joins other transactions to form a block, which then appends to the existing blockchain in chronological order. This technology is based on several key concepts. Decentralisation ensures equal ownership of data for everyone in the network. This, combined with the complete visibility of all transactions, guarantees unparalleled transparency. Additionally, the inherent nature of blockchain ensures that once an entry is made, it is permanent and safe from tampering (He and Turner 2022). Furthermore, cryptographic functions make blockchains resistant to breaches, making them secure data fortresses.

Various types of blockchains have emerged, including public chains such as Bitcoin, private variants such as Hyperledger, and consortium models such as R3, each catering to distinct needs and access levels. While its association with cryptocurrencies such as Bitcoin is prominent, the potential and utility of blockchain extend far beyond. In the contemporary era, the importance of blockchain is accentuated by its promise of decentralisation, offering freedom from centralised authority and guaranteeing transparency and security. Its ability to bypass intermediaries in various processes underscores its potential for greater efficiency and cost reduction. Innovations such as smart contracts and emerging business models, such as Decentralised Finance (DeFi) and Non-fungible Tokens, further attest to its transformative potential (Wan *et al* 2022). As industries across the spectrum gravitate towards it, blockchain's growth projections and investments soar, pointing to a promising decentralised future.

4.2 Literature review

4.2.1 Examining common factors for successful blockchain implementation

Although rife with potential, the complex world of blockchain technology demands an intricate understanding and comprehensive planning for a successful deployment. Both academic and industry experts have identified key determinants pivotal to successful blockchain adoption. In particular, the fundamental choice between public, private, or consortium blockchains is pivotal. For example, the open nature of public blockchains such as Bitcoin contrasts with the restricted access of private blockchains, making the latter more appropriate for certain business solutions (Makani *et al* 2022). Equally essential is the decision between permissioned and permissionless designs, which dictates who has the authority to validate transactions.

Scalability is a recurrent concern in blockchain dialogues. Effective implementations must address the transactional volume and speed that the organisation demands, juxtaposing decentralisation with optimal performance (Xu *et al* 2022). Although the blockchain boasts robust security, it is not impervious to threats. Aspects such as private key management and the looming threat of a 51% attack warrant attention, particularly in Proof-of-Work (PoW) systems (Xu and He 2022). As the digital world evolves towards interconnectedness, a blockchain's capacity to seamlessly mesh with other systems gains importance. Emphasising this, Xu *et al* (2022) discussed the significance of cross-chain communication and the harmonisation of blockchain solutions with existing legacy systems.

The clarity and structure of governance in a blockchain system cannot be understated. Decision-making protocols, including validation processes and protocol updates, are cornerstones that influence the system's credibility and functionality (Saari *et al* 2022). The quest for blockchain implementation is resource intensive. In addition, Saari *et al* (2022) found that ensuring a roster of skilled blockchain developers and equipping them with the right hardware and software tools can be challenging. Finally, one cannot stress enough the importance of a granular understanding of blockchain's myriad facets. Comprehending its principles, from distributed ledgers to consensus mechanisms, is imperative. Völter *et al* (2021) cautioned against venturing into this domain without adequate knowledge. In summary, while blockchain's transformative potential spans sectors, its effective deployment hinges on a gamut of carefully considered factors. This insight, drawn from the literature, seeks to guide those navigating the challenging, yet promising, waters of blockchain technology.

4.2.2 Legal challenges and considerations for blockchain implementation and adoption

Blockchain, an innovation touted for its revolutionary potential, is not exempt from legal scrutiny and challenges. Its decentralised design, while promising also unravels a myriad of legal concerns that any prospective adopter should consider. A prime challenge comes from regulatory uncertainty. Due to blockchain's nascent status, many countries grapple with its effective regulation. For example, the SEC in the United States is still navigating the intricate task of classifying and regulating diverse cryptocurrencies operating on blockchain platforms (Ferrández-Pastor *et al* 2022). This decentralised technology, allowing swift cross-border transactions, raises jurisdictional issues. The conundrum is to determine which nation's laws apply when disputes emerge from multi-country blockchain transactions (Krichen *et al* 2022).

Data protection regulations, such as the General Data Protection Regulation (GDPR) and California Consumer Privacy Act (CCPA), highlight the privacy challenges blockchain encounters. Notably, blockchain's immutability is at odds with rights such as the 'right to be forgotten.' Once information finds its way into a blockchain, erasing or altering it is not straightforward, potentially clashing with extant data protection laws (Gad *et al* 2022). The allure of smart contracts is undeniable. However, their insertion into the legal ecosystem is not seamless. How they dovetail with classical contract law and their interpretation in litigious situations remains an area of debate (Lemos *et al* 2022). Intellectual property rights, too, face a turbulent relationship with the blockchain. Technology can offer indisputable proof of ownership, but the convoluted landscape of intellectual property laws, varying by jurisdiction, adds complexity. What one nation views as acceptable use might be perceived as a copyright infringement in another (Habib *et al* 2022).

Companies opting for blockchain technology in finance face stringent financial regulations. Given their inherent design, cryptocurrency platforms find adhering to Anti Money Laundering (AML) and Know Your Customer (KYC) norms a formidable challenge (Mahmudnia *et al* 2022). Decentralisation also instigates liability issues. In scenarios where a blockchain failure inflicts losses on a user, pinning responsibility, whether on the platform creator, the user, or the network's participants, is intricate (Saldamli *et al* 2022). Finally, licensing and operational prerequisites vary between industries. For example, a firm that wishes to leverage blockchain for banking might need specific licences and adhere to stipulated operational norms, including capital sufficiency and risk containment (Abdelmaboud *et al* 2022). In essence, while blockchain promises transformation across sectors, its integration into the legal framework necessitates careful navigation, informed by both technological understanding and legal expertise.

4.3 Methodology: bibliometric analysis

Bibliometric analysis is a statistical methodology that offers a unique lens through which one can analyse and quantify the growth, patterns, and impacts of scientific publications within a particular domain. Applying this methodology, we aim to dive deeper into the scholarly landscape of blockchain-related publications. Initially, a comprehensive search of notable academic databases, principally Scopus will be carried out to extract all blockchain-related publications, from their inception until the present. These data will then be categorised annually, allowing us to chart the growth trend of these publications over time. Through the data extraction process, we will also capture the affiliation and country of each author. By visualising this data on a map, we aim to depict the intensity of research activity across regions, offering insights into the countries most actively contributing to blockchain research, especially when considering applications by industry. Figure 4.1 shows the data extraction. Figures 4.2, 4.3, and 4.4 show a distribution of publications per author, co-occurrence of keywords, and publications by country, respectively.

Another crucial aspect to consider is citation data. By examining how frequently publications and authors are cited, we can rank and highlight those that have profoundly impacted the field. Furthermore, gathering information on which papers

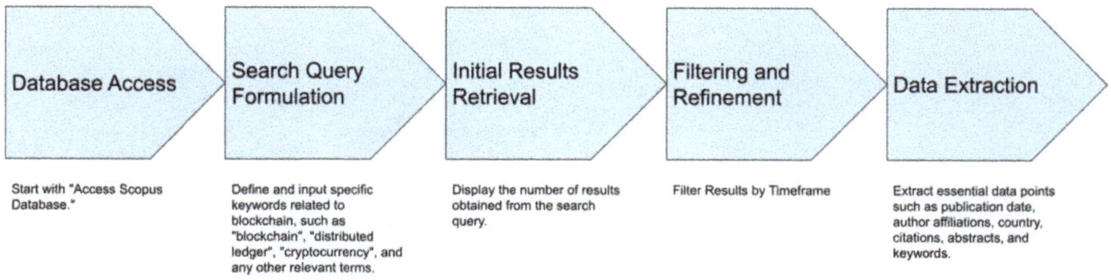

Figure 4.1. Process of data extraction.

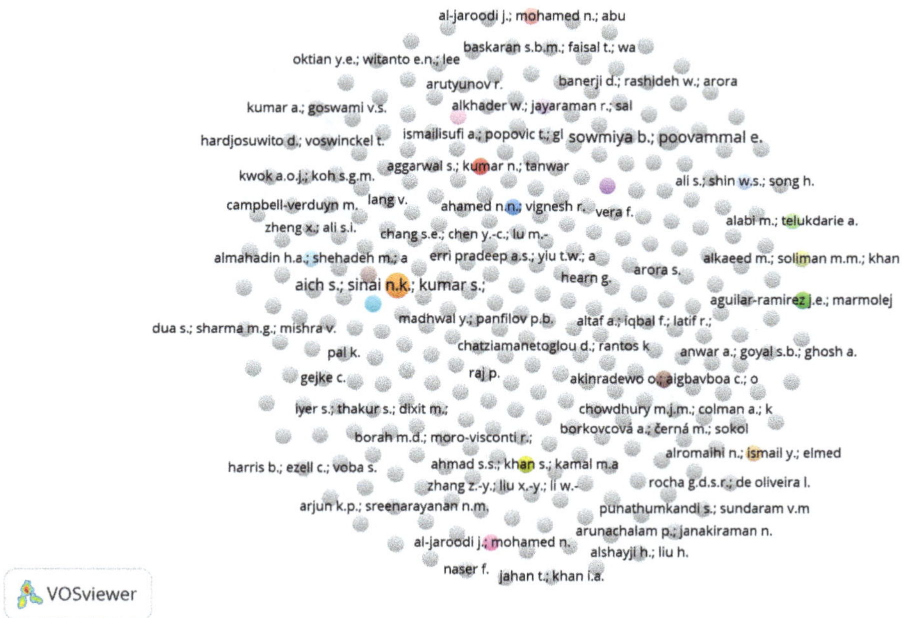

Figure 4.2. Publications per author.

frequently cite others can construct a co-citation network using network analysis tools. Such an approach will reveal tight-knit clusters of research papers, suggesting specific thematic concentrations, and providing clarity on the primary areas of blockchain research and their interrelated nature. Finally, to better understand the existing literature's themes and gaps, we will extract abstracts and keywords from the gathered publications. The content will be analysed to identify prevalent topics, emerging themes, and any potential under-researched areas using text mining and natural language processing techniques. By harnessing the power of bibliometric analysis, this research aims to illuminate the trajectory and current trends in blockchain studies. Such insights will undoubtedly pave the way for future research endeavours, catalysing further advancements in the field.

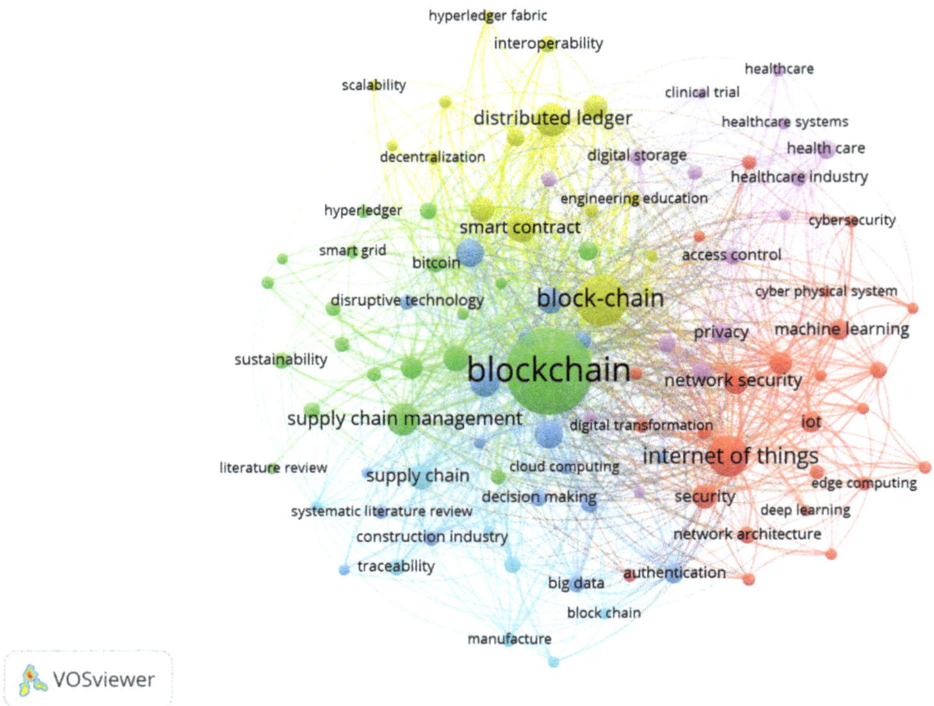

Figure 4.3. Co-occurrence keywords.

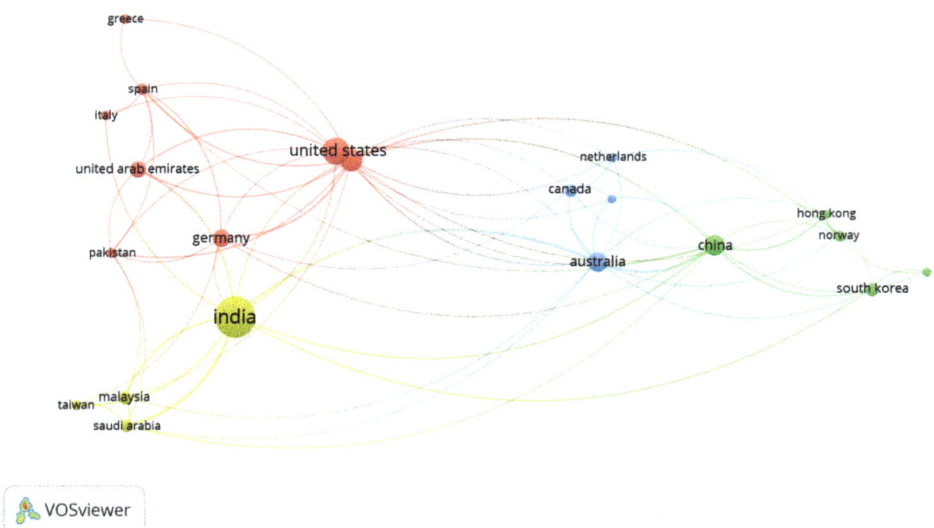

Figure 4.4. Publications per country.

4.4 Discussion

4.4.1 Applications in various industries

Blockchain technology, a decentralised and distributed ledger system, is making inroads into a wide array of industries, fundamentally reshaping their operational dynamics. Figure 4.5 shows various fields of application of blockchain. In the education sector, one of the most profound applications of blockchain lies in credential verification. The system can store and validate academic credentials, ensuring their authenticity and integrity (Chen *et al* 2022). Such an application can substantially reduce academic fraud, giving employers a more efficient mechanism to authenticate the qualifications of potential hires.

Financial services is arguably one of the industries most affected by blockchain technology. Cryptocurrencies such as Bitcoin and Ethereum are founded on blockchain principles, providing decentralised financial assets. Moreover, blockchain's ability to enable direct peer-to-peer (P2P) transactions is transformative because it eliminates the need for intermediaries, which often results in cost savings and enhanced efficiency. This is particularly crucial for processes such as money transfers, remittances, and general payments. Furthermore, the advent of smart contracts, which are self-executing contracts with the terms directly written into lines of code, can potentially revolutionise how financial agreements are made. These contracts streamline processes, reduce the risk of manual errors, and offer more

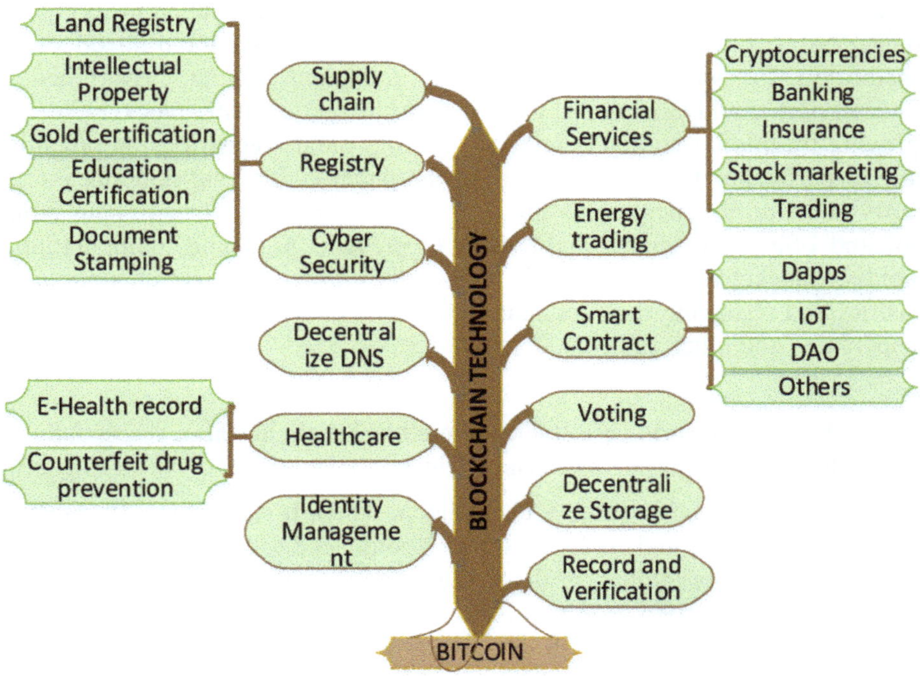

Figure 4.5. Blockchain application in various sectors. Reprinted from Sanka *et al* (2021), Copyright (2021), with permission from Elsevier.

transparency (Ghosh *et al* 2023). Additionally, DeFi has opened a new chapter in the financial world, promoting P2P financial services while sidelining traditional intermediaries.

Supply chain management will benefit enormously from blockchain integration. With the ability to provide unparalleled transparency and traceability, the blockchain can meticulously track the journey of goods from their origin to the consumer. This ensures product authenticity, deters counterfeiting, and instills more trust in the end consumer. Several progressive companies have already embarked on their blockchain journey in their supply chains to reap these benefits. Healthcare, an industry known for its fragmented data systems, finds solace in blockchain's promise of a unified, secure, and interoperable data system (Mahmudnia *et al* 2022). Be it managing patient records with enhanced security or ensuring traceability in clinical trials and drug production, blockchain appears as a beacon of transformation.

The real estate sector is buzzing with the potent of blockchain. From property transactions to land registries, blockchain can offer transparency, diminish fraud, and simplify the transaction process by sidelining intermediaries. Smart contracts in this domain can potentially automate rental and sales agreements, marking a significant leap from traditional systems. In the energy sector, blockchains herald a new era, especially for renewable sources. P2P energy trading could be mainstreamed, facilitating the growth of decentralised renewable energy grids (Krichen *et al* 2022). Furthermore, blockchain has the ability to refine grid management practices and ensure better transparency in renewable energy certificates.

Media and entertainment find value in blockchain for intellectual property rights management. It offers innovative avenues for royalty tracking and introduces new monetisation strategies, especially in the digital content domain. Government and public services could undergo a massive transformation with blockchain. From improving traditional voting systems to securing public registries, the potential is great. In particular, blockchain could be the panacea for long-standing issues of security and transparency for voting systems, ensuring that votes remain tamper-proof and offering verifiable audit trails to prevent fraud and manipulation. In summary, blockchain's potential is vast and multifaceted, permeating various industries and reshaping their foundational processes with the promise of security, transparency, and efficiency.

4.4.2 Analysing real-world case studies where blockchain has been successfully implemented in various industrial settings

The University of Nicosia faced a challenge in which it was difficult to verify the authenticity of educational credentials, which could lead to employment fraud. As a solution, they became the first in the world to publish academic certificates on the blockchain. This provided a decentralised method to verify the authenticity of the credentials, making it easier for employers and other institutions to validate academic credentials and reduce fraud instantly. Similarly, the University of Bahrain grappled with slow and inconvenient traditional certification systems for graduates and employers (Saari *et al* 2022). By issuing digital diplomas on the blockchain, they streamlined the verification process for employers, effectively reducing administrative burdens and practically eliminating fraud.

In the financial sector, traditional cross-border payments are known to be expensive and time-consuming. JPMorgan Chase addressed this issue by developing JPM Coin, a digital currency to enable instant payment transfers and simplify the funds transfer process. As a result, transactions became faster, more affordable, and more efficient. Santander Bank tackled the inefficiencies in traditional bond issuance processes, which were complex, expensive and required several intermediaries. In September 2019, they launched the world's first end-to-end blockchain bond using the public Ethereum blockchain for issuance (Mahmudnia *et al* 2022). This move helped reduce intermediaries, accelerate the issuance process, and increase transparency.

In supply chain management, Walmart and IBM partnered to address the complexity and lack of transparency in the food industry. They created Food Trust, a blockchain-based system to trace food items from farm to store. This not only enhanced transparency but also enabled faster responses to food safety issues. The diamond industry, represented by De Beers, has been plagued by issues surrounding conflict diamonds, fraud, and a non-transparent supply chain. De Beers introduced a blockchain platform called 'Tracr' to log the journey of diamonds from extraction to retail, ensuring that each diamond's history was permanently and immutably recorded (Krichen *et al* 2022). This initiative bolstered transparency and consumer confidence in ethically sourced diamonds.

Maersk, the world's largest shipping company, identified problems within the logistics and shipping sector, primarily the overwhelming paperwork that could introduce errors and delays. Collaborating with IBM, they developed TradeLens, a platform that digitises the supply chain process, streamlining paperwork and enhancing transparency. In healthcare, the scattering of patient medical records across various providers makes data access difficult. MedRec, built on the Ethereum blockchain, offers patients an immutable log of their medical information, improving data access, management, and security. Similarly, Estonia recognised the need to secure health data and empowered patients to control data access (Saari *et al* 2022). They developed a blockchain system that significantly improved data security, making Estonia a pioneer in digital healthcare solutions.

Propy, a global property store, addressed the complexities and potential fraud in real estate transactions by leveraging the Ethereum blockchain. It utilises smart contracts to execute property transactions, increasing transparency and security in the sector. Sweden's land registry, Lantmäteriet, also moved toward blockchain to make property transactions more efficient, transparent, and secure. In the energy sector, Grid Singularity unveiled D3A software and a decentralised energy data exchange platform that uses blockchain technology to enhance energy data transparency and promote better grid management. Greeneum uses blockchain to track carbon credits and renewable energy certificates, integrating AI for analytics (Mahmudnia *et al* 2022). This has fortified the carbon credit market and stabilised energy prices.

Ecocoin rewards sustainable actions and tracks carbon footprints using blockchain, promoting sustainable behaviour. Meanwhile, Solarcoin gives tokens to solar energy producers, incentivising the adoption of solar energy. Share&Charge connects electric vehicle owners with charging stations, automating payments, while Bankymoon provides blockchain-based energy meters and cryptocurrency utility payments. LO3 Energy and Power Ledger both focus on decentralising energy grids, making renewable energy trading more efficient. In the entertainment sector, Mycelia, initiated by musician Imogen Heap, uses blockchain to ensure artists are compensated fairly. Similarly, SingularDTV provides a platform for artists to directly distribute content to consumers (Krichen *et al* 2022). Spotify, recognising challenges in royalty payments and copyright management, acquired blockchain startup Mediachain to connect artists and tracks for more efficient royalty distribution.

Estonia's move to secure digital identities and offer e-residency using blockchain has made digital services more trustworthy and efficient. Dubai's government aims to leverage blockchain for all government documents, anticipating substantial time and cost savings. The Plastic Bank encourages people in impoverished areas to collect and recycle plastic waste, offering digital tokens in return. West Virginia introduced blockchain in its voting system, ensuring integrity and accessibility, especially for overseas voters. Voatz offers a blockchain-backed mobile voting solution that promotes transparency and security. Lemonade, an insurtech firm, overcomes the inefficiencies in traditional insurance claims processing using blockchain and AI, speeding up the process considerably. IOTA addresses the increasing need for a secure and scalable solution for machine-to-machine transactions in the Internet of Things with its unique Tangle blockchain, paving the way for new business models and ensuring data integrity.

4.4.3 Comparing and contrasting the use of blockchain across various industries, assessing the impact and effectiveness in each case

4.4.3.1 Financial services vs. healthcare

The decentralised and transparent nature of blockchain is intrinsically adaptable for the financial services industry. Cryptocurrencies such as Bitcoin offer a vision of an alternative monetary system outside traditional banking structures. Moreover, major banking institutions such as JPMorgan are diving into this realm; with the introduction of the JPM Coin, there is clear evidence of the potential for traditional finance to meld with blockchain technologies. However, when we move to healthcare, the scenario changes. The sector's priorities lean heavily towards patient data confidentiality, integrity, and cross-platform operability. Blockchain undeniably holds promise in meeting these demands, but its real-world application in health has been noticeably gradual, primarily due to the stringent regulatory landscape and the imperative nature of data privacy (Saari *et al* 2022). Projects such as MIT's MedRec, which promotes a decentralised system for managing patient data, illuminate blockchain's potential adaptability in this sector.

The financial sector's embrace of blockchain has been widespread and enthusiastic. From multinational banking giants to emerging fintech startups, many entities have integrated blockchain into their operations, leading to services such as instantaneous cross-border transactions, smart contracts, and the emerging field of DeFi. In contrast, healthcare's dance with blockchain is still in its early stages, largely limited to pilot projects and research explorations. Initiatives are currently geared toward streamlining patient data management, ensuring drug traceability and increasing transparency in clinical trials. Regarding immediate impact, the finance world has benefited from tangible rewards from blockchain integration. Enhanced transaction speeds, reduced costs, robust security measures, and the birth of innovative financial instruments underscore this transformation. However, healthcare's gains from blockchain are seen as long-term investments (Mahmudnia *et al* 2022). The potential here is vast, with improved patient data systems leading to better patient care, increased research capabilities, and a more personalised healthcare experience.

The path to blockchain integration is, however, full of hurdles. For the financial world, these include regulatory ambiguities, the need for extensive industry consensus (particularly vital for multibank transactions), and scalability and energy consumption concerns. Healthcare faces its own set of unique challenges, including navigating stringent data privacy mandates such as the US's HIPAA (Health Insurance Portability and Accountability Act) and Europe's GDPR, devising health record systems that are both secure and user-friendly, and the intricate task of blending blockchain with preexisting healthcare IT setups. Peering into the future, blockchain's potential in financial services seems vast and varied, with possibilities ranging from a wholly decentralised financial landscape to advanced global remittance methods and asset tokenisation. For healthcare, the prospects are equally exciting, including universally accessible health records, groundbreaking improvements in research and clinical trials, and a paradigm where data security and patient privacy are paramount.

4.4.3.2 Supply chain management vs. government or public services

The unparalleled transparency and unalterable nature of the blockchain find a compelling application in supply chain management. This technology offers an unbroken and tamper-resistant record of products as they travel through their supply chain. Such an approach increases traceability and combats fraudulent activities. The collaboration between IBM and Maersk, resulting in the TradeLens solution, is a testament to this, because it presents a radical transformation in the shipping domain by providing complete visibility across the supply chain. In contrast, the government or public services sector presents a mosaic of requirements, spanning from managing data and facilitating services to streamlining voting and identity verification mechanisms. Here, blockchain's adaptability exhibits a richer texture. The case of the West Virginia mobile voting pilot, which leveraged blockchain, stands as a marker of the potential for enhanced transparency and diminished fraud (Krichen *et al* 2022). However, reservations persist about the security and the ability to scale such solutions.

Taking into account the scale of adoption, the supply chain sector displays a burgeoning embrace of the blockchain. Multiple global conglomerates and logistics entities are either already harnessing this technology or are in advanced stages of exploration. This paints a vibrant picture of extensive blockchain integration in the supply chain landscape. However, the narrative changes when one shifts the lens to the government or public services sector. Blockchain's footprint here is developing and exhibits stark variances, contingent on the region and the service in focus. While countries such as Estonia emerge as vanguards in the integration of blockchain into public services, the global spectrum needs to be more diverse in its adoption. The ripple effects and advantages are unmistakable in supply chain management. Blockchain can improve efficiency, curtail expenses, improve product traceability, and prevent fraud. However, the influence is more overarching within the government and public services domain. Here, it harbours the potential to make operations more transparent, pare down bureaucratic red tape, refine data handling methods, and uplift the quality of services provided to citizens (Saari *et al* 2022). In particular, while the benefits in the supply chain sphere are immediate and palpable, the government sector reaps the rewards in a more long-term and subtle manner.

The journey towards blockchain integration is, however, full of obstacles. For supply chain management, hurdles manifest in syncing with prevailing systems, crafting industry-standard norms, and striking a harmonious balance between transparency and preservation of privacy. The government and public services sector grapples with challenges, most notably data privacy, navigating the regulatory maze, fostering widespread public sector engagement, and contending with potential pushback given blockchain's decentralised ethos. Projecting into the future, blockchain's use in supply chain management appears to be very promising. The prospects encompass the creation of entirely transparent supply chains, the problem of counterfeit products, and even the development of novel business archetypes such as decentralised trading platforms. At the same time, the governmental and public services arena also is on the cusp of transformative potential. The blockchain could metamorphose aspects such as managing public records, electoral systems, digital identification mechanisms, and public services delivery mechanisms.

4.4.3.3 Energy vs. real estate

Blockchain's unique attributes of decentralisation, transparency, and enhanced security have found resonance in various industries, notably energy and real estate, each with its own set of challenges and potential. In the energy sector, blockchain's characteristics make it an ideal solution to tackle issues such as managing distributed energy resources, facilitating P2P energy trade, and meticulously tracking carbon emissions. An embodiment of this is the Brooklyn Microgrid project, which harnesses blockchain's prowess to craft a community-driven energy-sharing ecosystem. However, the embrace of blockchain in this domain is still nascent, largely characterised by small-scale pilot ventures or research-driven initiatives. Despite this, some pioneering utility entities and startups are exploring the avenues that this technology can unveil. While the promise is immense, the actual benefits remain unrealised. The vision for the energy sector is to leverage blockchain to create decentralised energy marketplaces, refine grid administration, and support sustainable energy practices. Yet, the path forward is full of impediments (Mahmudnia *et al* 2022). The sector faces regulatory snags, technical complications related to the orchestration of energy resources on the blockchain, and the imperative for broader participation of the ecosystem.

The real estate industry presents a different narrative. Here, blockchain's adaptability manifests itself in processes such as property transactions, meticulous title management, and the innovative concept of fractional ownership. Platforms such as Propy stand as testaments to this, having seamlessly executed cross-border property deals via blockchain, hinting at a profound paradigm shift in traditional real estate dealings. Unlike its counterpart in the energy sector, blockchain's implementation in real estate is more profound. Startups and established real estate firms are venturing into the blockchain for property deals, maintaining title records, and the emerging realm of tokenisation. The fruits of this integration are more immediate and palpable. Blockchain introduces greater transparency, diminishes the risk of fraudulent activities, streamlines transactions, and potentially democratises real estate investment via tokenisation. However, this sector is not insulated from challenges (Krichen *et al* 2022). It grapples with regulatory complexities, the intricacies of integrating with legacy systems, the urgency for broader industry endorsement, and navigating the nuanced legal tapestry related to property ownership interwoven with blockchain.

Looking into the future, the canvas of possibilities is expansive. The energy sector with blockchain as its ally, can envisage a landscape where energy markets are not only decentralised but also democratised and sustainable energy practices are not just a vision but a reality. Currently, in the real estate arena, blockchain holds the promise of globalising property markets, popularising fractional real estate ownership, and pioneering smart real estate contracts, potentially reshaping the contours of the industry.

4.4.3.4 Media and entertainment vs. education

The transformative potential of blockchain technology has been recognised in many diverse sectors, with the media, entertainment, and education industries as notable examples. Each industry, with its unique needs, presents distinct opportunities and challenges when dealing with blockchain.

In the realm of media and entertainment, the decentralised, transparent, and programmable nature of blockchain makes it poised to address perennial challenges, notably in the facets of digital rights verification, equitable revenue dissemination, and forging novel content distribution paradigms. Take, for example, Audius, a pioneering platform that leverages blockchain to orchestrate a more egalitarian and decentralised music streaming ecosystem. Although the enthusiasm is palpable, the broad-scale adoption of blockchain in this sector is still budding. Both emerging startups and seasoned players are tapping their toes into the waters of digital rights management, content dissemination, and meticulous royalty oversight via blockchain. The expansive vision here is democratising content access, ensuring creators receive their due, and injecting unparalleled transparency into content ownership and distribution channels. However, the fruition of these transformative benefits are still unrealised (Mahmudnia *et al* 2022). The path forward has many hurdles as the industry grapples with regulatory problems, the imperative for pan-industry embracement, and the intricate dance of marrying blockchain with entrenched digital rights and distribution frameworks.

Similarly, the education sector discerns blockchain as an ally to navigate its unique challenges. Technology promises solutions such as impregnable credentialing mechanisms, infallible verification processes for academic laurels, and the empowerment of learners with autonomous record management. Sony Global Education's foray into blockchain, yielding a robust system for the secure dissemination of student records, stands as a testament to these possibilities. Yet, much like its media counterpart, the education sector's blockchain adoption storey is still being written, and is characterised by sporadic experimentation by avant-garde universities and institutions. The prospective rewards are attractive, hinting at a future with tamper-proof credentials, learners wielding unprecedented control over their academic narratives, and a marked uptick in administrative prowess (Krichen *et al* 2022). However, the path is strewn with many challenges, particularly in navigating the intricate maze of data privacy, ensuring seamless interoperability across disparate institutional systems, and aligning with regulatory frameworks and standardisation imperatives.

Looking into the future, the vistas opened by blockchain are vast and varied. Decentralised content ecosystems, equitable creator remuneration structures, and a transparent regime for digital rights oversight could redefine the media and entertainment sphere. Simultaneously, the education sector could witness revolutionary changes, with blockchain underpinning digital accreditation, fostering lifelong learning trajectories, and reshaping the very anatomy of student data stewardship.

4.4.4 Evaluating the benefits and drawbacks of using blockchain technology in different industries

4.4.4.1 Education

Education, an industry synonymous with the moulding of futures and the bridging of gaps, is entering a phase of evolution with the advent of blockchain technology. The intersection of these two realms is replete with both thrilling possibilities and thought-provoking challenges. On the optimistic side of the spectrum, blockchain's inherent characteristics present tantalising benefits to the education sector. The first among these is the virtue of transparency and verifiability. The scourge of degree fraud, which has long engulfed educational institutions and employers, finds a formidable adversary in the blockchain. This technology offers an indisputable mechanism to authenticate qualifications, thereby streamlining the often onerous verification processes that institutions undertake and ensuring employers' trust in the credibility of academic credentials (Saari *et al* 2022). Furthermore, the sanctity of blockchain immutability ensures a bedrock of trust; once committed to the chain, a record is inviolable against tampering or illicit alteration.

Blockchain extends an empowering promise to learners moving deeper into the personal realm. It offers students unparalleled dominion over their educational narratives, allowing them to arbitrate access and maintain their privacy. No longer are students beholden to institutional bureaucracies to share or access their records; they become the stewards of their academic data. In tandem with this, blockchain also paves the way for a holistic learning portfolio. Beyond the confines of formal education, learners can now weave a diverse tapestry of their educational journey, from MOOCs (Massive Open Online Courses) to informal learning episodes, professional advancement courses, and other eclectic experiences, all under one consolidated and verifiable course. Yet, the horizon is not without its clouds. The journey to integrate blockchain into education is fraught with significant challenges. One of the most overarching is the huge task of adoption and standardisation. The global educational arena is a vast mosaic of institutions, each with its established protocols and systems. Orchestrating a consensus among these multiple entities to embrace a single blockchain protocol is no mean feat.

Furthermore, despite the robustness of encryption, the spectre of privacy concerns is large. Inscribing sensitive personal and academic information on a blockchain could inadvertently expose vulnerabilities, or at least stir fears about potential data breaches. Finally, the practicality of technology adoption also raises eyebrows. The educational landscape is diverse, with institutions ranging from ivy-clad universities to modest community schools. The technological prowess and resources at their disposal are equally varied, potentially leading to disparities in blockchain adoption and implementation. In conclusion, as the education sector grapples with the growing potential of blockchain, it stands at a crossroads of transformative promise and pragmatic challenges. The road ahead is undeniably exciting, but it requires thoughtful navigation to truly harness the potential of this symbiosis.

4.4.4.2 Financial services

The financial services sector, often considered the backbone of the global economy, has witnessed a plethora of changes over the years, and the emergence of blockchain technology is heralding another transformative phase. Just as the advantages of intertwining blockchain with financial services are numerous, so are the intricacies that accompany this innovative amalgamation. Blockchain presents tantalising prospects by peering through the lens of benefits. One of its standout merits is the enhancement of efficiency and transparency. By design, blockchain negates the need for intermediaries, optimising transaction processes. This can significantly truncate transaction times, slashing costs, and bolstering the overall efficiency of financial exchanges (Krichen *et al* 2022). Simultaneously, the transparency afforded by blockchain's immutable ledger system substantially curtails the avenues for financial deception or fraud.

Beyond domestic borders, the realm of cross-border transactions is undergoing a metamorphosis thanks to the blockchain. Historically, these transactions have been burdened with bureaucratic red tape, sluggish processing times, and exorbitant fees. Blockchain can overhaul this domain, making transnational financial exchanges faster, more cost-effective, and fundamentally streamlined. But perhaps the most avant-garde benefit arising from the marriage of blockchain and finance is the sprouting of innovative financial services. DeFi embodies this revolution (Saari *et al* 2022). Bypassing traditional banking systems, DeFi platforms harness blockchain to offer financial instruments ranging from loans to interest-bearing accounts, democratising finance and making it more accessible to the masses.

However, juxtaposed with these compelling advantages is a set of intricate challenges. The realm of finance is intricately meshed with regulations and the integration of blockchain treads on uncertain regulatory grounds. This lack of clarity and the potential legal ambiguities act as deterrents for many financial entities, causing trepidation in the adoption of the whole blockchain. On the technical front, scalability emerges as a pronounced concern. Financial transactions, especially in bustling economic hubs, occur at an astonishing velocity and volume (Mahmudnia *et al* 2022). Today's blockchain networks, especially those that leverage PoW, are grappling with processing these transactions with the alacrity that the financial sector demands.

Finally, the volatility associated with cryptocurrencies, arguably blockchain's poster child in the financial world, cannot be sidelined. The wild price swings inherent to Bitcoin and Ethereum cast a shadow over their viability as stable financial instruments, potentially undermining the broader acceptance of blockchain in finance. In the grand tapestry of financial services, blockchain technology paints a picture that is equal parts promise and complexity. As the financial world teeters on the brink of this new era, it must weigh the transformative potential of blockchain against the challenges it presents.

4.4.4.3 Supply chain management

Supply chain management (SCM) stands as a cornerstone for numerous businesses worldwide. With globalisation, supply chains have morphed into vast and intricate networks. Ensuring efficiency, transparency, and authenticity throughout this network is paramount, and this is where the role of blockchain comes to the fore. However, like every technological marvel, the fusion of blockchain and SCM presents both advantages and challenges. From the benefits, the potential for traceability is profound. In traditional systems, tracking a product back to its origin could be a daunting task, riddled with inaccuracies. Blockchain technology offers businesses the ability to provide visibility from start to finish. This not only ensures better quality control, but is also invaluable in substantiating the genuineness of products (Krichen *et al* 2022). This traceability is vital for industries such as food and pharmaceuticals, where the source of ingredients can be a matter of life and death.

Closely linked to traceability is the boon of transparency. Blockchain, with its decentralised ledger system, minimises the shadowy corners where fraudulent activities or errors might lurk. This increased transparency can be a game changer, especially in complex supply chains. It can lead to improved inventory management, fewer counterfeit products, and a higher trust quotient among stakeholders. Yet another feather in the cap of blockchain's benefits is the concept of smart contracts. These self-executing contracts with the terms of agreement directly written into lines of code can automate many processes within the supply chain (Ghosh *et al* 2023). From ensuring timely payments to cementing agreements without the need for intermediaries, smart contracts can be the linchpin for improved efficiency.

On the flip side, challenges are also inherent. One of the key points is the potential for data entry errors. The axiom 'garbage in, garbage out' holds true for blockchains. Its indelible nature means that any mistake made at the data entry phase becomes solidified, potentially causing ripple effects throughout the supply chain. The challenges do not end there. For blockchain to truly revolutionise SCM, widespread adoption is required. This means that not just one entity but all players within a supply chain, from raw material providers to retailers, need to be on the blockchain bandwagon. Furthermore, the interoperability issue looms large (Abdelmaboud *et al* 2022). Given the plethora of blockchain platforms, crafting systems that seamlessly interact with multiple blockchains is a task that is easier said than done.

Finally, the very process of integrating the blockchain into SCM is neither simple nor cheap. The labyrinthine nature of supply chains coupled with the intricacies of blockchain technology can make the implementation complex. The capital required might be prohibitive, especially for smaller enterprises operating on thin margins. In essence, while blockchain promises to reshape and refine supply chain management, it is a journey fraught with both opportunities and obstacles. As businesses navigate this path, a judicious balance between taking advantage of the benefits and mitigating the challenges will be crucial.

4.4.4.4 Healthcare

The healthcare sector is at the nexus of human well-being and technological advancement. A primary concern in healthcare has always been the management, security, and accessibility of patient data. With the onset of blockchain technology, the industry is witnessing the potential for a transformative change. However, the journey is as much about challenges as it is about opportunities. In terms of benefits, one of the most significant is the potential for data security and interoperability. Current systems, which are characterised by siloed databases and varying standards, are rife with vulnerabilities. Blockchain could usher in an era where data security is substantially enhanced. The shared immutable history ensures that every transaction is traceable and irreversible, minimising the chances of data breaches or unauthorised alterations (Saldamli *et al* 2022). This, coupled with the potential for better interoperability, could greatly enhance the efficiency and security of health record management.

Drug traceability stands out as another promising application of blockchain. The pharmaceutical industry, a critical part of healthcare care, grapples with the issue of counterfeit drugs that pose risks to patient safety and industry credibility. Blockchain could provide a transparent and unalterable ledger of drug origin and movement, ensuring that only authentic medications reach end consumers. Another advantage that blockchain offers is the shift towards a patient-centric approach. Instead of being passive recipients, patients can have the reins of their health records, deciding who gets access and when (Chen *et al* 2022). This not only empowers patients but also ensures greater data accuracy because patients can verify and validate their information.

However, the path to blockchain integration in healthcare care is laden with challenges. One of the primary hurdles is the establishment of data standards. Healthcare data are multifaceted, spanning everything from clinical records to imaging to genomic data. Creating a universally accepted standard for encoding and storing these data on blockchain platforms is undeniably daunting. Furthermore, the healthcare sector is heavily bound by regulations. Compliance with frameworks such as HIPAA and GDPR is essential (Mahmudnia *et al* 2022). With blockchain still in its nascent stages in healthcare care, there are grey areas regarding how data storage and sharing would align with these regulatory mandates.

Finally, the sheer task of implementing blockchain in healthcare cannot be underestimated. Transitioning from legacy systems to blockchain is not about flipping a switch—it requires substantial investment, both in terms of capital and time, and given the vastness and complexity of the healthcare sector, the challenges are magnified. To conclude, while blockchain holds tremendous potential to revolutionise the healthcare landscape, it is imperative to tread with caution and clarity (Habib *et al* 2022). The balance between leveraging transformative benefits and navigating inherent challenges will be pivotal in the successful adoption of blockchain in healthcare.

4.4.4.5 Real estate

The real estate sector, which is historically characterised by its complexity, lengthy procedures, and the involvement of numerous intermediaries, is now at the forefront of industries that could benefit enormously from the integration of blockchain technology. This technology promises not only to address several pain points in the industry but also to introduce innovations that could transform the way in which real estate transactions are conducted. However, as with any new technology's integration, blockchain's adoption in real estate comes with its own set of challenges. One of the most prominent benefits that blockchain brings to the table is transparency and efficiency. The traditional real estate process often involves numerous checks, balances, and verifications. With the introduction of the blockchain, each transaction can be recorded on an immutable ledger, providing a clear and tamper-proof history of property ownership (Lemos *et al* 2022). This transparency can substantially reduce the risk of fraud, and make the transaction process more streamlined and efficient.

The concept of tokenisation in real estate is another game changer. Through this, real estate assets can be divided into tokens that represent a fraction of the property. This allows for greater liquidity in the market because these tokens can be easily traded or sold. Furthermore, it allows more people to invest in real estate by purchasing tokens rather than entire properties. In addition, the introduction of smart contracts could potentially overhaul the real estate transaction process. These contracts, which are self-executing with the terms directly written into code, can automate various aspects of the transaction, such as the transfer of property titles and funds (Gad *et al* 2022). This could substantially reduce the need for intermediaries, such as escrow services, thereby speeding up transactions and reducing associated costs.

On the flip side, regulatory compliance is one of the primary challenges facing the integration of blockchain in real estate. Given that real estate regulations vary significantly from one jurisdiction to another, creating a one-size-fits-all blockchain solution is arduous. Each solution would need to take into account local laws and practices, which adds to the complexity. Furthermore, the adoption challenge cannot be overlooked. Real estate stakeholders, many of whom have been embedded in traditional systems and methodologies for years, may be resistant to adopting a new technology (Krichen *et al* 2022). This resistance can stem from a lack of understanding, fear of obsolescence, or simply comfort with the known methods.

Finally, the issue of market volatility, especially when dealing with real estate-backed cryptocurrencies, is pertinent. The real estate market, which is traditionally viewed as stable, could be introduced into the volatile world of cryptocurrencies, adding a layer of unpredictability. In conclusion, while the blockchain undoubtedly has the potential to revolutionise the real estate sector, its integration needs to be thoughtful and strategic. Balancing the immense benefits while navigating the challenges will be the key to a successful transformation.

4.4.4.6 Energy

The energy sector, with its vast infrastructure and intricate systems, is one of the pillars of modern civilisation. However, with increasing demands for sustainability, decentralisation, and transparency, the industry needs innovative solutions. With its inherent attributes, blockchain technology has emerged as a potential catalyst for change in this sector. While the benefits are substantial, the integration of blockchain into energy systems comes with a set of complexities. One of the most groundbreaking benefits blockchain offers the energy sector is the potential for P2P energy trading. In the traditional model, energy production is centralised, typically in the hands of large utilities. But with the rise of decentralised renewable energy sources such as solar panels and wind turbines, the number of smaller producers is growing. Blockchain can allow these producers to trade or sell their excess energy directly to consumers (Ferrández-Pastor *et al* 2022). This democratises the energy market, possibly leading to more competitive prices and promoting renewable energy adoption.

In addition, the nature of blockchain ensures transparency and efficiency in grid management. All transactions, whether related to energy trading or other aspects, are securely recorded in an immutable ledger. This can reduce administrative overhead, minimise errors, and ensure that every transaction is transparent to all participants. Furthermore, the push towards green and sustainable energy solutions has necessitated the introduction of renewable energy certificates (Völter *et al* 2021). Blockchain can ensure that these certificates, which attest to the energy produced from renewable sources, are issued, tracked, and traded securely, eliminating the chances of fraud.

However, the path to integrating blockchain into the energy sector is fraught with challenges. One of the significant hurdles is the regulatory environment. The energy sector across the globe is strictly regulated, given its strategic importance. Introducing a decentralised model, especially one that could disrupt the current power dynamics, would require not just technological innovations but also substantial policy adjustments. Scalability and efficiency are also concerns. The energy sector operates on a massive scale, with millions of transactions happening simultaneously. Presently, popular blockchain systems such as Bitcoin or Ethereum cannot handle such transaction volumes efficiently (Saari *et al* 2022). While there are ongoing efforts to improve blockchain scalability, this remains a concern.

Finally, the integration of reliable digital identity systems is crucial. There must be foolproof identity verification mechanisms to ensure secure and seamless transactions on a blockchain-based energy platform. The development and integration of these systems can be complex. In summary, while blockchain has immense potential to redefine the energy sector, its practical implementation demands a balanced approach. The industry needs to weigh the undeniable benefits against the challenges and adapt accordingly (Xu and He 2022). If done correctly, the blockchain could indeed herald a new era of decentralised, transparent, and sustainable energy systems.

4.4.4.7 Media and entertainment

Media and entertainment are the core pillars of contemporary culture, driving recreation and the dissemination of ideas, values, and narratives. The advent of the internet and digital platforms has drastically altered this, for better and worse. With its multifaceted benefits, blockchain has the potential to solve some of the challenges digital media faces while further enhancing its virtues. One of the most pressing challenges in the digital media space is digital rights management (DRM). The ease with which digital content can be replicated and distributed has made piracy a rampant problem. Blockchain can be a game changer here. Creating an immutable record of ownership and rights can ensure that content creators get their rightful royalties and prevent unauthorised distribution (Xu and He 2022). Furthermore, through smart contracts, payments can be automated and recorded transparently, ensuring that artists and creators are fairly compensated.

Another advantage of blockchain in this space is the democratisation of the media landscape. Historically, artists and creators had to rely on intermediaries such as studios, publishers, and record labels to reach their audience. These intermediaries often took a significant cut from the profits. Blockchain allows for direct artist-audience interaction, ensuring that a larger share of the revenue goes directly to the creators. Moreover, blockchain's ability to verify authenticity can be vital in an age of rampant piracy and counterfeit content (Xu *et al* 2022). Blockchain can help ensure that consumers get genuine content and that the creators' intellectual property rights are safeguarded.

However, this bright potential is not devoid of challenges. The first of these is the regulatory landscape. Intellectual property laws vary dramatically between regions. Creating a blockchain solution that adheres to the diverse set of rules while maintaining universality is complex. Scalability is another challenge. The media and entertainment industry deals with vast data and transactions (Makani *et al* 2022). The leading current blockchain platforms might not be equipped to handle this volume efficiently, especially in real-time scenarios like streaming.

Finally, for blockchain to be truly transformative in this space, it needs widespread adoption. This requires artists, consumers, and industry players to understand, trust, and embrace technology, which can be a slow process given the novelty of blockchain and the conservative nature of some industry stakeholders. In conclusion, while blockchain has the potential to redefine the media and entertainment industry, its practical application will require careful navigation of the challenges (Wan *et al* 2022). Blockchain could lead to a more equitable, transparent, and enriched media landscape if harnessed correctly.

4.4.4.8 Government and public services

Governments worldwide are tasked with immense responsibility, from managing public funds to ensuring the welfare of their citizens. The nature of government functions demands the highest degree of trust, efficiency, and accountability. In recent years, blockchain has emerged as a potential tool that can revolutionise how governments operate, offering benefits that can optimise various aspects of governance. However, as with any new technology, its adoption in the public sector is not without challenges. One of the core strengths of blockchain is its ability to offer transparency and efficiency. By integrating blockchain into public processes, governments can significantly reduce bureaucracy and paperwork, making administrative processes smoother and faster. This has the potential to save both time and money. For instance, blockchain can help streamline land registration, licence renewals, and even tax collection (He and Turner 2022). providing an immutable record of transactions also reduces the chances of fraud and misallocation of public funds.

Another notable benefit of blockchain in governance is its role in enhancing citizen engagement. One of the most promising use cases is digital voting. By providing a secure and tamper-proof system for electronic voting, blockchain can make electoral processes more transparent and accessible. Additionally, blockchain-based digital identity systems can simplify citizen verification for various public services, reducing fraudulent claims and ensuring that benefits reach the intended recipients. However, while the potential advantages are numerous, the challenges of blockchain adoption in governance cannot be ignored. Regulatory implications are one of the foremost concerns. Integrating blockchain into public services requires a comprehensive legal framework to ensure its ethical and secure use (Renduchintala *et al* 2022). Given the decentralised nature of blockchain, data protection becomes a significant challenge, especially in regions with stringent data privacy laws.

Public acceptance is another obstacle. While tech enthusiasts might be eager to embrace blockchain, the broader public may be hesitant, primarily due to a lack of understanding of the technology. Misconceptions and concerns about blockchain's association with cryptocurrencies might further hinder its acceptance. Finally, the security and privacy concerns associated with using blockchain for government functions are paramount. While blockchain is secure by design, it is not impervious to attacks or flaws. Given the sensitive nature of the data that governments handle, any breach can have catastrophic implications. In conclusion, while blockchain has the potential to transform governance and public services, its adoption should be approached with caution and thorough planning (Sunny *et al* 2022). Collaboration between technology experts, policymakers, and stakeholders will be crucial in ensuring that technology integration is both effective and secure.

4.5 Summary

Blockchain technology has progressively presented itself as a revolutionary tool with vast potential across diverse sectors, from healthcare and real estate to government and media. Its inherent qualities of transparency, immutability, and decentralisation offer innovative solutions to age-old challenges, promoting efficiency, trust, and direct engagements. However, like any transformative technology, it is accompanied by challenges such as scalability, regulatory uncertainty, and the need for broad public understanding and acceptance. As sectors grapple with the decision to incorporate blockchain, it becomes evident that a balanced approach is crucial. Although the benefits are promising, the potential pitfalls require careful attention. Successful integration will require collaboration between industry experts, regulators, and stakeholders to ensure that the technology is harnessed to its full potential while addressing associated risks. Ultimately, the evolving journey of blockchain's integration into our societal systems will undoubtedly be a testament to our collective adaptability, innovation, and vision for a more transparent and efficient future.

References

Abdelmaboud A, Ahmed A I A, Abaker M, Eisa T A E, Albasheer H, Ghorashi S A and Karim F K 2022 Blockchain for IoT applications: taxonomy, platforms, recent advances, challenges and future research directions *Electronics* **11** 630

Chen Y, Lu Y, Bulysheva L and Kataev M Y 2022 Applications of Blockchain in Industry 4.0: a review *Inf. Syst. Front.* pp 1–15

Ferrández-Pastor F-J, Mora-Pascual J and Díaz-Lajara D 2022 Agricultural traceability model based on IoT and blockchain: application in industrial hemp production *J. Ind. Inform. Integr.* **29** 100381

Gad A G, Mosa D T, Abualigah L and Abohany A A 2022 Emerging trends in blockchain technology and applications: a review and outlook *J. King Saud Univ.—Comput. Inform. Sci.* **34**

Ghosh P K, Chakraborty A, Hasan M, Rashid K and Siddique A H 2023 Blockchain application in healthcare systems: a review *Systems* **11** 38

Habib G, Sharma S, Ibrahim S, Ahmad I, Qureshi S and Ishfaq M 2022 Blockchain technology: benefits, challenges, applications, and integration of blockchain technology with cloud computing *Future Internet* **14** 341

He Z and Turner P 2022 Blockchain applications in forestry: a systematic literature review *Appl. Sci.* **12** 3723

Krichen M, Ammi M, Mihoub A and Almutiq M 2022 Blockchain for modern applications: a survey *Sensors* **22** 5274

Lemos C, Ramos R F, Moro S and Oliveira P M 2022 Stick or twist—the rise of blockchain applications in marketing management *Sustainability* **14** 4172

Mahmudnia D, Arashpour M and Yang R 2022 Blockchain in construction management: applications, advantages and limitations *Autom. Constr.* **140** 104379

Makani S, Pittala R, Alsayed E, Aloqaily M and Jararweh Y 2022 A survey of blockchain applications in sustainable and smart cities *Cluster Comput.* **25** 3915–36

Renduchintala T, Alfauri H, Yang Z, Di Pietro R and Jain R 2022 A survey of blockchain applications in the FinTech sector *J. Open Innov.: Technol., Market, Complex.* **8** 185

Saldamli G, Upadhyay C, Jadhav D, Shrishrimal R, Patil B and Tawalbeh L 2022 Improved gossip protocol for blockchain applications *Cluster Comput.* **25** 1915–26

Sanka A I, Irfan M, Huang I and Cheung R C 2021 A survey of breakthrough in blockchain technology: adoptions, applications, challenges and future research *Comput. Commun.* **169** 179–201

Saari A, Vimpari J and Junnila S 2022 Blockchain in real estate: Recent developments and empirical applications *Land Use Policy* **121**

Sunny F A, Hajek P, Munk M, Abedin M Z, Satu M S, Efat M I A and Islam M J 2022 A systematic review of blockchain applications *IEEE Access* **10** 59155–77

Völter F, Urbach N and Padget J 2021 Trusting the trust machine: evaluating trust signals of blockchain applications *Int. J. Inf. Manage.* 102429

Wan Y, Gao Y and Hu Y 2022 Blockchain application and collaborative innovation in the manufacturing industry: based on the perspective of social trust *Technol. Forecast. Soc. Change* **177** 121540

Xu X and He Y 2022 Blockchain application in modern logistics information sharing: a review and case study analysis *Prod. Plan. Control* 1–15

Xu X, Tatge L, Xu X and Liu Y 2022 Blockchain applications in the supply chain management in German automotive industry *Prod. Plan. Control* 1–15

Nnamdi Nwulu and Uyikumhe Damisa

Chapter 5

Drivers of blockchain adoption in the energy sector

The global energy sector is experiencing significant transformation, driven primarily by the imperatives of climate change and technological advancements. Central to this evolution is blockchain technology, which is a decentralised, transparent, and secure system initially designed for cryptocurrency transactions. This chapter delves into the multifaceted applications of blockchain in various energy-related areas, including mitigating climate change, fostering the adoption of renewable energy, and modernising power grids. Blockchain's attributes, including its ability to facilitate peer-to-peer (P2P) transactions, provide transparency, and automate processes through smart contracts, are uniquely suited to address the challenges and capitalise on the opportunities in the energy sector. As this chapter discusses, the marriage of blockchain and energy could revolutionise operations, enhance efficiency, empower consumers, and fortify the path toward a sustainable future. However, to harness the full potential of this convergence, the integration challenges of technology, regulations, and stakeholder collaboration must be robustly addressed.

5.1 Introduction

Blockchain technology, which is a distributed ledger system, has emerged as a revolutionary tool in the digital age, offering a way to record transactions across multiple computers so that any recorded data becomes almost unalterable without altering all subsequent blocks. This technology first came into the limelight with the advent of Bitcoin after the 2008 financial crisis. Beyond its role in underpinning cryptocurrencies, blockchain has proven its potential across various sectors, including finance, which supports cryptocurrencies and smart contracts, and healthcare, facilitating secure and transparent record keeping. In the global landscape, the

doi:10.1088/978-0-7503-6295-5ch5

energy sector is critical, powering everything from homes to businesses, and entire industries. However, it faces many challenges, including decentralising energy resources, integrating renewable energy sources, ensuring transparent transactions in energy trading, and addressing security concerns (Vairagade and Brahmananda 2021). Recognising these challenges brought about a noticeable shift toward the digitalisation of the energy sector, aiming for greater efficiency, scalability, and adaptability.

Interestingly, the core attributes of blockchain, such as decentralisation, transparency, and immutability, align well with the challenges faced by the energy sector. This alignment has resulted in numerous pilot projects around the world exploring blockchain application in energy markets. Examples range from P2P energy trading platforms to renewable energy certificates, highlighting the growing intersection of these two domains. Understanding the drivers behind the adoption of blockchain in the energy sector is crucial, given the wide range of stakeholders involved. These drivers encompass the promise of improved transaction transparency and efficiency, the potential for decentralised energy resource management, the facilitation of renewable energy source integration, increased security and fraud prevention, and the creation of innovative business models and services (Swain *et al* 2022). As this exploration continues, the transformative potential at the intersection of blockchain technology and the energy sector will undoubtedly be a subject of keen interest and rigorous study.

5.2 Climate change

Climate change, a multifaceted environmental crisis, is the result of long-term changes in temperature and weather patterns, primarily caused by human activities. This phenomenon, rooted in the science of our atmosphere and the greenhouse gases it contains, has far-reaching impacts. Environmentally, climate change leads to melting glaciers, rising sea levels, and altered weather patterns. Economically, the consequences are felt in the agriculture, insurance, and infrastructure sectors. From a social perspective, it displaces populations, alters livelihoods, and can exacerbate inequality. Central to the climate change discussion is the energy sector, which has seen evolving consumption patterns over the past several decades. Historically, energy consumption revolved around wood and biomass. However, with the Industrial Revolution and subsequent developments, fossil fuels, such as coal, oil, and natural gas, became dominant. These energy sources, while effective, have been significant contributors to greenhouse gas emissions (Dzobo *et al* 2021). A statistical analysis reveals that the emissions of the energy sector surpass most other sectors, making it a primary target for climate mitigation.

Recognising the role of the energy sector, there has been a global push towards sustainable energy solutions, including the rise of renewables such as solar, wind, and hydro. Efforts have also been made to improve energy efficiency and develop technologies such as carbon capture and storage. On the international front, the Paris Agreement stands as a testament to global commitment, wherein nations have submitted their national determined contributions to reduce emissions (Roth *et al*

2022). However, transitioning to a sustainable energy paradigm is fraught with challenges. There are technological barriers to integrating renewables, economic challenges in financing green technologies, and political hurdles due to divergent interests.

In this matrix, blockchain technology emerges as a potential game changer. At its core, the blockchain is a distributed ledger system that offers transparent and immutable transaction records. In the realm of climate change mitigation, its applications are varied. Blockchain can streamline carbon credit trading, ensuring transparency and eliminating issues such as double counting. It also offers transparency in the supply chain, ensuring sustainable practices from the source to shelf. In the energy sector, in particular, blockchain is expected to revolutionise operations. By enabling P2P energy trading, it can promote decentralisation, reduce transmission losses, and improve efficiency. Tokenising energy assets can drive investments in renewables, while crowdfunding platforms can finance green projects. In addition, in the quest for energy efficiency, the need for efficient grid management is paramount. As demand response and load balancing become crucial, the blockchain provides real-time data sharing, and its smart contracts can automate demand responses, optimising grid operations. However, blockchain integration is not without challenges (Yang *et al* 2021). Technological limitations, regulatory and security concerns, and stakeholder buy-in must be addressed for this digital tool to be a true catalyst for climate mitigation.

5.3 Adoption of renewable energy

The current global energy scenario underscores an urgent shift towards renewable sources. Renewable energy and distributed energy resources (DERs) are at the forefront of this transition. Unlike centralised energy systems, which focus on large-scale energy production and distribution, DERs decentralise energy production, bringing generation sources closer to consumption points. This decentralisation is pivotal for mitigating climate change, improving energy security, and ensuring energy independence. DERs encompass a range of technologies. Solar panels, including photovoltaics, tap into the sun's energy. Wind turbines, especially small-scale and vertical-axis variants, harness wind power. Energy storage solutions, such as lithium-ion batteries and pumped storage, store energy for off-peak usage. Combined heat and power systems cogenerate heat and electricity, maximising efficiency (Nizamuddin and Abugabah 2021). Demand response tools, such as smart thermostats and advanced metering infrastructures, also provide more control over energy consumption.

Embracing DERs has many benefits; for example, they reduce transmission losses, offer economic savings to consumers and communities, democratise energy production, enhance grid reliability, and present significant environmental advantages by reducing carbon footprints and other emissions. However, the implementation of DERs is not without challenges. Grid integration can be complex and there are technical issues to overcome, such as ensuring grid stability and addressing energy storage concerns. The initial costs of installing these systems can be high, and there is a need to ensure a favourable return on investment. The variable nature of some renewables, such as solar and wind, raises concerns about reliability (Wu *et al* 2022). Additionally, regulatory and policy barriers often slow down DER adoption.

This is where blockchain, a groundbreaking technology, enters the equation, potentially accelerating DER adoption. At its essence, the blockchain is a decentralised ledger system that ensures transparent and immutable transactions. Blockchain can improve transparency, facilitate real-time transactions, and even streamline direct energy transactions within the energy sector, bypassing traditional utility structures. This paves the way for 'prosumers', i.e., consumers who produce and consume energy. Blockchain-powered platforms can allow for real-time energy trading, consumer empowerment, and the creation of local energy markets. For example, P2P energy trading platforms have emerged globally, enabling local energy transactions. In New York, the Brooklyn Microgrid allowed residents to sell excess solar energy to their neighbours using a blockchain platform, demonstrating the feasibility of localised energy markets (Delardas and Giannos 2022). Similarly, projects in Bangkok, Augsburg, South Korea, Bangalore, and Kyoto have showcased blockchain's potential in facilitating local energy solutions and trade.

Beyond P2P trading, the blockchain can also play a pivotal role in grid management. The integration of more DERs necessitates agile grid management. Blockchain can help maintain grid stability, with smart contracts allowing for automated demand-response measures, grid balancing, real-time monitoring, and even predictive maintenance. The transparency features further enable the verification of the green credentials of energy sources, boosting consumer trust (Teodorescu and Korchagina 2021). Token-based systems can verify and reward renewable energy production, promote green investments, and even streamline the trade of carbon credits.

The financial mechanisms are also getting a boost. Platforms such as MyBit, Jumpstart Energy Transfer, WePower, and SolarCoin utilise blockchain to democratise investments in DER projects, tokenise renewable energy certificates, or incentivise solar energy production. The management of the national and regional grid has similarly benefited. In places such as Singapore, Spain, the Netherlands, the UK, and Germany, blockchain has facilitated the integration of various energy sources into grids, improved energy trading transparency, and introduced mechanisms for real-time grid stabilisation. In summary, as the world steers towards renewable energy, blockchain stands out as a transformative tool, ensuring that the shift is efficient, transparent, and inclusive (Aysan *et al* 2021). The synergy between renewable energy and blockchain offers promising solutions to some of the most pressing challenges in the energy sector today.

5.4 Decentralised modern power grid

The evolution of modern power grids has brought new challenges and opportunities. As the global energy landscape transforms, decentralisation stands out as a pivotal trend, with blockchain technology playing a critical role. Historically, the conventional power grid model revolved around large centralised energy generation, which transmits electricity over vast distances to end consumers. Today, this model is gravitating towards more localised and adaptable energy systems, evolving in response to technological advancements and the dynamic needs of consumers. Decentralisation in the energy sector represents a shift from this centralised energy generation to a more distributed form, where power is produced much closer to its consumption point (Salama *et al* 2023). Here, blockchain, a decentralised ledger technology, comes to the forefront because it facilitates transparent, secure, and direct transactions, which eliminates the need for intermediaries.

The world's rising population and rapid urbanisation exert increasing pressure on traditional grids, highlighting their inefficiencies and vulnerabilities. Ageing infrastructure is often ill equipped to handle this growing demand. Moreover, as urban areas grow, there is an escalating requirement for reliable energy solutions. DERs, such as home solar installations, community wind farms, and energy storage solutions, emerge as localised energy sources to address these needs. Decentralised systems offer multiple advantages, including redundancy, resilience, and consumer empowerment. By having many energy sources, they ensure that a failure in one does not cripple the entire system. Furthermore, decentralisation allow consumers to transition into 'prosumers,' simultaneously producing and consuming their energy (Khezami *et al* 2022). By generating energy closer to where it is consumed, we also see a notable reduction in transmission losses.

This is where blockchain can amplify the benefits of decentralisation. Beyond being a trust-creating system with its immutable ledgers and distributed databases, blockchain introduces the concept of smart contracts, automated contracts executed through code. Such a system ensures that energy transactions are transparent and tamper-proof. It also makes P2P energy trading feasible, enabling consumers to trade excess energy directly. By minimising middlemen, transaction costs are further reduced. When it comes to implementing these decentralised power grids with blockchain, several models stand out. For example, local microgrids are autonomous energy grids that can operate independently of the primary grid. An example of this in action is the Brooklyn Microgrid, where residents trade solar energy using a blockchain. On a broader scale, integrating diverse energy sources into the grid can be simplified with blockchain by tokenising energy assets and incentivising green energy through blockchain-based rewards (Kaur *et al* 2022). Power Ledger exemplifies this potential by focusing on the integration of various energy sources.

Nationally and regionally, the impact is profound. It can improve grid reliability and adaptability, efficiently managing the integration of myriad energy sources. Additionally, it smooths the complexities involved in assimilating large-scale renewable sources into the grid. The efforts of Electron in the UK illustrate how blockchain can revolutionise national grid operations. In essence, the symbiosis of decentralisation and blockchain in the energy sector signifies more than a technological evolution (Saraji and Khalaf 2022). It heralds a movement towards a sustainable, efficient, and consumer-driven energy future.

5.5 The emergence of prosumers

The term 'prosumer' has gained prominence in the evolving energy landscape. A combination of 'producer' and 'consumer,' prosumers in the energy sector refer to individuals or entities that both produce and consume energy. Historically, energy was generated by large centralised entities, and consumers simply purchased it for use. However, with technological advancements and a global emphasis on sustainability, individuals are now empowered to produce their own energy, especially from renewable sources. This transformation has been made possible by the decentralisation of energy systems and the rise of DERs, such as solar panels and community energy projects. This shift promises energy independence, reliability, and notable environmental benefits (Swain *et al* 2022). On the economic front, the consumer model can be lucrative, allowing individuals to sell surplus energy either back to the grid or directly to neighbours.

Blockchain technology has emerged as a critical facilitator for the consumer movement. Its decentralised nature promotes P2P energy trading, reducing the need for intermediaries and ensuring transparent and trustful transactions. One stand-out feature of blockchain in this domain is the smart contract, which facilitates automated energy transactions based on prevailing rates. Additionally, the ability to tokenise energy transforms it into a digital asset that can be traded or rewarded, thus further incentivising green energy endeavours. Several real-world implementations exemplify the synergy of blockchain and prosumers. Energi Mine in the UK has created a platform that rewards energy-saving actions, promoting a proactive approach among consumers. Meanwhile, SunContract in Slovenia uses blockchain for direct P2P energy trading, connecting consumers with local renewable energy producers. Australia's Myco platform prioritises homeowners with solar setups, ensuring that they receive optimal returns for surplus energy. Elektra in Mexico uses blockchain to counter energy poverty, enabling remote communities to access energy resources efficiently. Wepower in Lithuania has devised a system that tokenises green energy, giving businesses an avenue to directly acquire green energy. Grid+ in the USA offers real-time energy pricing, encouraging consumers to make informed choices. Conjoule in Germany has created a blockchain-based local energy marketplace, fostering direct sales between consumers and neighbours. Finally, ElaadNL in the Netherlands integrates electric vehicles with the grid using blockchain, ensuring seamless charging transactions. These innovative endeavours showcase the transformative potential when decentralised energy production converges with blockchain (Dzobo *et al* 2021). The rise of prosumers underscores a significant paradigm shift in the energy world, pointing to a future where people champion sustainability, efficiency, and economic progression.

5.6 Pooling or sharing of resources

The modern energy landscape is increasingly recognising the importance of optimising the use of energy resources. Inefficiencies, lack of transparency, and adaptability constraints have hampered traditional approaches to resource pooling and sharing. This is where blockchain technology, known for its decentralised ledger capabilities, offers a promising solution to the energy sector's challenges. Historically, energy grids have been centralised, leading to challenges in demand-response management and difficulties during peak load times. With their inherent intermittency, the rise of renewable energy sources further accentuates the need for better resource management (Vairagade and Brahmananda 2021). While microgrids and DERs offer a solution, the importance of optimal use of capacity cannot be overstated for economic and environmental reasons.

Blockchain technology emerges as a beacon of hope in this context. It promises decentralisation, transparent and tamper-proof records, and the ability to create smart contracts that can automate complex agreements. Through blockchain, the concept of P2P energy trading becomes a reality, enabling real-time tracking of energy demand and supply without intermediaries. This not only reduces costs but also streamlines energy transactions. Several real-world implementations underscore the potential of blockchain in the energy sector. For example, a project in the UK by Verv used artificial intelligence (AI) and blockchain to predict and facilitate energy trading among residents of a London housing estate. This initiative highlighted the potential of community-driven and decentralised energy markets. Similarly, the Energy Web Foundation introduced the Energy Web Chain to harmonise energy assets of varying scales, ensuring efficient use (Swain *et al* 2022). Groupe Renault's collaboration with The Mobility House in France showcased the dual potential of electric vehicles (EVs) in a smart charging programme, emphasising the vehicles' roles as both energy consumers and suppliers.

In Japan, the Gavagai platform has allowed residents to trade surplus solar energy with their neighbours, optimising renewable energy usage within communities. Germany's GreenCom Networks connected diverse home energy devices, transforming individual homes into interconnected hubs of energy exchange. Another German initiative, Share&Charge, leveraged blockchain to allow individuals to share their private EV charging stations, maximising infrastructure usage (Dzobo *et al* 2021). In essence, these pioneering projects provide a window into a future where technology and innovation combine to offer a more efficient, inclusive, and sustainable approach to energy resource management.

5.7 Summary

The landscape of the global energy sector is undergoing a transformative change. Anchored by increasing concerns over climate change, there is a compelling drive towards more sustainable, efficient, and decentralised energy solutions. In this transformative journey, blockchain technology emerges as a formidable ally. Its decentralised, transparent, and secure features align seamlessly with the evolving aspirations of the energy sector. From facilitating cryptocurrency transactions to enabling P2P energy trades, blockchain's versatility is evident. The synergy between blockchain technology and the energy sector offers a promising avenue to address some of the most pressing challenges of today. Whether it is the integration of DERs, the transparency in green energy credentials, or the modernisation of power grids, blockchain seems poised to play a pivotal role. It not only provides technological solutions and empowers consumers but it also fosters an environment of trust and inclusivity.

Furthermore, the ability of blockchain to democratise energy markets, tokenise energy assets, and provide real-time data sharing places it at the epicentre of the sector's digital evolution. As the world gravitates towards a more sustainable energy paradigm, the convergence of blockchain and energy promises to be one of our times' most intriguing and impactful collaborations. However, the full potential of this synergy will only be realised when challenges such as technological limitations, regulatory frameworks, and stakeholder collaboration are effectively addressed. As the narrative unfolds, stakeholders, from policymakers to industry leaders and consumers, must engage in collective action to harness the potential of this dynamic intersection.

References

Aysan A F, Bergigui F and Disli M 2021 Blockchain-based solutions in achieving SDGs after COVID-19 *J. Open Innov.: Technol., Market, Complex.* **7** 151

Delardas O and Giannos P 2022 Towards energy transition: use of blockchain in renewable certificates to support sustainability commitments *Sustainability* **15** 258

Dzobo O, Malila B and Sithole L 2021 Proposed framework for blockchain technology in a decentralised energy network *Protect. Control Modern Power Syst.* **6** 1–11

Kaur N, Bhullar S and Seth N 2022 Application of blockchain technology in sustainable energy systems *Blockchain Technology in Corporate Governance: Transforming Business and Industries* (New York: Wiley) 143–61

Khezami N, Gharbi N, Neji B and Braiek N B 2022 Blockchain technology implementation in the energy sector: comprehensive literature review and mapping *Sustainability* **14** 15826

Nizamuddin N and Abugabah A 2021 Blockchain for automotive: an insight towards the IPFS blockchain-based auto insurance sector *Int. J. Electr. Comput. Eng. (IJECE* **11** 2443

Roth T, Utz M, Baumgarte F, Rieger A, Sedlmeir J and Strüker J 2022 Electricity powered by blockchain: a review with a European perspective *Appl. Energy* **325** 119799

Salama R, Al-Turjman F, Altrjman C, Kumar S and Chaudhary P 2023 A Comprehensive survey of blockchain-powered cybersecurity—a survey *IEEE Xplore.* 774–7

Saraji S and Khalaf C 2022 Blockchain applications in the energy industry. www.igi-global.com; IGI Global (https://igi-global.com/chapter/blockchain-applications-in-the-energy-industry/287690)

Swain A, Swain K P, Palai G and Mohanty M N 2022 *A Blockchain-Powered Energy Monitoring System* (Springer eBooks) pp 253–72

Teodorescu M and Korchagina E 2021 Applying blockchain in the modern supply chain management: its implication on open innovation *J. Open Innov.: Technol. Mark., Complex* **7** 80

Vairagade R S and Brahmananda S H 2021 Enabling machine learning-based side-chaining for improving QoS in blockchain-powered IoT networks *Trans. Emerg. Telecommun. Technol.* **33** e4433

Wu Y, Wu Y, Cimen H, Vasquez J C and Guerrero J M 2022 Towards collective energy community: potential roles of microgrid and blockchain to go beyond P2P energy trading *Appl. Energy* **314** 119003

Yang Q, Zhou D, Yiheng C, Zhang X and Su Y 2021 Evolve from traditional power grid network to blockchain and iot integrated network *IEEE Xplore.* 119–24

Chapter 6

Blockchain's integration in the energy industry: practical implications

Integrating blockchain technology within the energy sector promises a transformative shift towards decentralisation, transparency, and enhanced transactional efficiency. However, this fusion presents a myriad of challenges encompassing technical hurdles, regulatory nuances, economic implications, and social considerations. This chapter delves into the intricacies of these challenges, from scalability and interoperability issues to stakeholder trust and environmental concerns. Through a comprehensive exploration, we emphasise the need for technological innovations, informed policy making, and collaborative efforts to navigate this intricate landscape. As the global energy paradigm evolves, understanding and addressing the complexities of the blockchain-energy nexus becomes imperative to forge a resilient, sustainable, and inclusive energy future.

6.1 Introduction

The blockchain can be understood as a digital ledger that records transactions on multiple decentralised nodes. This technology made its appearance with the rise of Bitcoin and cryptocurrencies. Unlike traditional centralised databases that are managed by single entities, blockchain operates on a peer-to-peer (P2P) network structure. In this system, nodes play a crucial role in the validation and recording of transactions, greatly contributing to its decentralised nature. This decentralisation has several implications, e.g., enhanced security because there is no single point of failure, increased transparency because every node has access to the complete ledger, and distributed trust where trust is disseminated across the network rather than being centralised. Delving deeper into its data structures, a block on the blockchain comprises a list of transactions detailing the sender, receiver, and value, a precise timestamp of its creation, and a unique cryptographic hash representing its content. This chain mechanism ensures continuity and security by linking blocks via their hashes, with each block containing the hash of its predecessor. The process of block creation and addition, known as mining, is computational and relies on consensus mechanisms such as Proof-of-Work (PoW) and Proof-of-Stake (PoS) to agree on the validity of transactions. One of the standout features of the blockchain is its immutability. The role of cryptographic hashing is paramount in this, where hash functions transform block content into a fixed-size code. Even a small change leads to a different hash, which ensures tamper resistance (Habib *et al* 2022). If one block is altered, then it disrupts the continuity of the subsequent blocks, requiring a high computational cost to change and ensuring that the majority consensus would reject tampering attempts.

Energy has always played an instrumental role in human civilisation, fuelling societal progress and technological advancements. Historically, before the industrial era, energy was primarily derived from human and animal labour, and basic renewable sources such as windmills and water wheels. The Industrial Revolution saw the rise of coal and the steam engine, driving industrial growth. The twentieth century marked the dominance of oil, the transformative influence on transportation, the electrification of cities, and the advent of nuclear energy. As we approached the late-twentieth and early-twenty-first century, there was a growing realisation of the limitations of fossil fuels, leading to the emergence of renewable energy sources, such as solar, wind, and hydroelectric power. Today, while fossil fuels such as oil, coal, and natural gas, continue to play a substantial role, renewables such as solar, wind, biomass, and hydroelectric are on the rise. Innovations in energy storage, distribution and advances in energy-efficient technologies have further shaped the industry. This industry is not just the backbone of transportation, housing, and urbanisation but is also the foundation of critical infrastructure such as hospitals, communications, utilities, and more (Maka and Alabid 2022). From an economic standpoint, energy drives growth, influences global trade, and is intertwined with daily household consumption.

The energy landscape is undergoing significant changes with trends such as decentralisation and digitalisation. This transformation is evident with the emergence of microgrids, renewable installations, and home energy systems, which contrast traditional centralised grids. Blockchain's introduction into this landscape allows for P2P energy trading, turning consumers into prosumers. However, the current centralised energy markets and intermediaries present challenges in tracking energy sources and transactions. Blockchain provides an immutable ledger solution that offers transparency and direct energy trading without intermediaries. Modern energy grids, while sophisticated, are susceptible to threats. These vulnerabilities, both digital (e.g. cyberattacks) and physical, are areas of concern (Reddy and Rao 2023). Blockchain can offer cryptographically secured transactions, adding a layer of trust and, with its decentralised consensus mechanisms, reduces single points of failure, which enhances grid resilience.

The intersection of blockchain and energy hints at evolving dynamics and early indications of blockchain's potential alignment with energy needs. Blockchain can streamline processes, reducing inefficiencies in centralised systems and challenges in integrating multiple energy sources. It offers transparent, traceable, and verifiable energy transactions, eliminating hidden costs and opaque supply chains. From a security point of view, the decentralised nature of blockchain ensures data integrity, resistance to tampering, and offers secure access control mechanisms (Rejeb *et al* 2023). Existing reviews on the blockchain-energy nexus might have limited cross-disciplinary analyses, a tendency to overemphasise potential over limitations, geographical and cultural biases, and could quickly become outdated given the rapid evolution of this technology.

The current dynamics of blockchain in the energy sector indicates that many past reviews may be outdated. With the evolution of blockchain technology and the energy landscape, there is a mismatch between past reviews and the current state. Previous reviews might have had a narrow focus, overlooking new applications or lacking cross-disciplinary insights (Valdivia and Balcell 2022). The surge in new research, technological breakthroughs, localised solutions, and the need for diverse voices underscores the importance of a comprehensive review.

Comprehensive reviews can significantly influence stakeholder decisions. Policymakers and regulators can derive insights for sound regulations, aligning them with technological advancements and fostering innovation. Industry leaders and entrepreneurs can gain strategic insights, guide investment decisions, and offer a competitive edge. For researchers, such reviews point out gaps, set agendas, and provide a foundation for new entrants (Ferrón-Vílchez *et al* 2020). Finally, consumers and end users can make informed energy choices, leading to proactive behaviours and enhanced trust.

The primary aim of this chapter is to present a comprehensive overview of blockchain's applications, challenges, and implications within the energy sector. This includes exploring specific applications, regulatory challenges, stakeholder viewpoints, and economic implications. The expected outcomes are a comprehensive understanding of the current state, a list of challenges, actionable recommendations, and insight into potential future developments.

6.2 Literature review

6.2.1 Blockchain evolution

Blockchain technology, in essence, is a decentralised, distributed, and decentralised ledger that provides unparalleled security, transparency, and immutability. The concept of decentralised systems rooted in cryptographic techniques, can be traced back to pioneering works such as those by Diffie and Hellman (1976). As these cryptographic techniques developed, the 1990s saw the rise of the Cypherpunk movement, which championed the causes of privacy and decentralisation. The game changer arrived with Satoshi Nakamoto's groundbreaking 2008 whitepaper titled 'Bitcoin: A Peer-to-Peer Electronic Cash System', (Nakamoto 2008) which redefined currency and ledgers. This was followed by the introduction of Litecoin in 2011, which promised faster transactions through a different hashing algorithm (George and Patatoukas 2020). Around the same time, Ripple emerged, which focuses on expedited cross-border transactions through a unique consensus mechanism.

Vitalik Buterin's Ethereum was introduced a few years later, which presented the world with the concept of a blockchain platform supporting smart contracts. This vision took a tangible form with Ethereum's launch in 2015 and the subsequent standardisation of tokens thanks to the ERC-20 standard. In the quest for privacy, blockchain platforms such as Monero and Zcash were introduced, which placed transaction privacy at their core. Third-generation platforms such as Cardano, with its research-centric methodology, and EOSIO, with its emphasis on scalability for decentralised application (DApp) development, signalled a continued evolution of the technology. The financial sector was not left untouched; platforms such as Compound, MakerDAO, and Aave started reimagining financial systems on the blockchain by 2018 (Bhutta *et al* 2021). Concurrently, Non-fungible Token (NFT) projects such as CryptoPunks and CryptoKitties have spotlighted the enormous potential of digital collectibles.

The consensus mechanisms also underwent innovation, with Peercoin's hybrid system leading the way in 2012. The importance of blockchain interoperability was soon recognised, and platforms such as Polkadot and Cosmos emphasised seamless multi-chain operations. Recent advances have been characterised by Ethereum's ambitious transition to Ethereum 2.0, Bitcoin's scalability and privacy improvements through the Taproot upgrade, and the emergence of layer-2 solutions such as Ethereum's Optimistic rollups and Bitcoin's Lightning Network (Gorkhali and Chowdhury 2021). Meanwhile, the DeFi landscape continues to grow, with platforms operating across multiple chains and the increasing prominence of stablecoins.

The NFT space is evolving, with a focus on utility and integration into metaverse platforms such as Decentraland. At the same time, solutions for decentralised identity, data storage, and real-time data streams have been developed, exemplified by projects such as Microsoft's ION, Filecoin, and IOTA's Tangle. In the cryptographic realm, the looming threat of quantum computing has accelerated research into post-quantum cryptographic techniques. Governance models are being redefined, as seen with the rise of decentralised autonomous organisations (DAOs) such as Uniswap's DAO (Krithika 2022). Given the increasing environmental concerns, initiatives such as Algorand's eco-friendly model and commitments to renewable energy usage are taking centre stage in the blockchain domain.

6.2.2 Blockchain research in the energy industry

Blockchain's unique qualities, such as transparency, security, and decentralisation, are driving innovative applications across the energy sector, ranging from renewables to conventional energy sources and from grid management to financing mechanisms. In the renewables domain, UK-based Electron has created platforms promoting shared access to flexible assets connected to the grid, exemplifying decentralised energy storage solutions (Gawusu *et al* 2022). Meanwhile, Energy Web Chain collaborates with wind energy producers to verify and document energy production, ensuring the environmental integrity of every kilowatt-hour produced.

In the oil and gas sector, Vopak from the Netherlands harnesses blockchain to enhance tank storage management efficiency. Meanwhile, PetroBloq from Canada offers solutions for the entire asset life cycle management to boost both efficiency and reliability. Grid management is another area that has seen blockchain applications. Grid Singularity in Austria uses this technology for dynamic pricing and demand response, facilitating real-time energy use adaptations based on grid requirements (Juszczyk and Shahzad 2022). In the United States, PJM Interconnection is looking to streamline interconnection processes for new producers using blockchain.

The integration of electric vehicles into the energy matrix is enhanced by platforms such as Germany's Share&Charge, which facilitates decentralised charging and blockchain-based payments. In addition, eMotorWerks in the United States uses blockchain to balance the energy flow between electric vehicles, the grid, and home energy solutions. Rosatom, Russia's state atomic energy corporation, is probing blockchain's potential to track nuclear fuel components to ensure authenticity. Carbon emissions monitoring is another pivotal application, with IBM's Green Horizon Project in China using blockchain to verify air quality data sources (Foti and Vavalis 2021). Veridium Labs in the United States employs blockchain to tokenise carbon credits, simplifying trade, and assuring each credit's authenticity.

Water management also benefits from blockchain. IBM and SweetSense have collaborated to monitor real-time groundwater usage in California, and Australia's Water Ledger uses it for real-time data management related to water rights and trading. Bioenergy applications include the BTL Group and Duncan, Lohn, and Leonard in Canada, piloting blockchain to authenticate and trace biofuel sources. In the realm of smart homes, IBM and Samsung's joint initiative, ADEPT, leverages blockchain to coordinate energy trades between smart appliances (Wang *et al* 2021). ME SOLshare in Bangladesh uses blockchain to enable rural households to trade surplus solar electricity.

On the financing front, WePower in Lithuania links green energy producers with investors via blockchain, which enables direct energy purchases at favourable rates. Acciona in Spain uses blockchain to trace renewable energy from hydropower sources, while GE Power in the United States partners with Tamr for blockchain-based data storage related to thermal power operations. In geothermal energy, Iceland's RENergy uses blockchain to verify the authenticity of the energy. In the hydrogen fuel sector, Power Ledger and Silicon Kingdom Holdings collaborate using blockchain to produce and trade green hydrogen (Parmentola *et al* 2021). South Korea's LG CNS has developed a blockchain solution to monitor battery life cycles.

EnergiMine in the UK uses blockchain to finance renewable projects through tokenised energy-saving credits. Gazprom in Russia is testing blockchain for efficient trade documentation in LNG and LPG trading. For district energy systems, Germany's Conjoule uses blockchain technology to track consumption and facilitate payments. Swytch in the United States harnesses blockchain to advise on energy efficiency and rewards sustainable behaviours with tokens (Gorkhali and Chowdhury 2021). Finally, cybersecurity for energy infrastructure gets a boost from

Estonia's Guardtime, which uses blockchain to ensure data transmission's integrity and origin.

6.3 Methodology

To ensure a comprehensive review of blockchain applications in the energy industry, a structured and systematic search strategy was implemented. This involved the use of databases such as IEEE Xplore, which is a leading platform for the technical literature in electrical engineering, computer science, and related areas. Google Scholar, a broad interdisciplinary search engine, was also employed, along with specific journals to ensure domain-specific insights. The search prioritised keywords such as 'Blockchain in energy,' 'Blockchain applications in renewables,' 'Decentralised energy systems,' 'Smart grid and blockchain,' 'Blockchain in oil and gas sector,' and 'Energy sector cybersecurity blockchain applications.' The search was narrowed down to articles and studies published between January 2021 and July 2023 to maintain relevance and capture the most recent developments.

For the selection criteria, peer-reviewed articles and papers that specifically focus on blockchain applications in the energy sector were prioritised. Special attention was paid to papers that provide empirical evidence or case studies. The review excluded non-English studies, papers that only tangentially mentioned blockchain without an in-depth exploration in the energy context, and articles without a clear energy applications focus. This strict selection was essential for a rigorous and focused review, allowing the inclusion of studies that delve deeply into blockchain's specific role in the energy sector.

Regarding data collection and analysis, a standardised form was used to extract relevant details from each study, such as objectives, methods, key findings, and implications. Thematic analysis grouped the data under common themes such as 'Blockchain in renewables' or 'Blockchain for grid management.' To ensure the reliability and validity of this approach, two independent reviewers reviewed the studies, ensuring consistent data interpretation. If discrepancies arose between reviewers, then they were resolved at consensus meetings. Finally, external domain experts were consulted to validate the findings and interpretations.

6.4 Discussion

6.4.1 The economic implications of blockchain integration within the energy sector

Analysing the economic implications of blockchain integration within the energy sector is essential to understand the transformative shift that this technology can introduce. Blockchain, a decentralised ledger system, holds promise in addressing various inefficiencies traditionally associated with the energy sector, from grid management to energy trading. The energy sector has historically navigated numerous challenges, from infrastructure costs to regulatory barriers. However, with the introduction and growth of blockchain technology, many of these challenges are now being addressed with innovative solutions. One of the key advantages is the potential for significant cost savings. Blockchain can improve operational efficiency by improving transparency and automating many operations that traditionally required manual intervention. The decentralised nature of blockchain reduces transaction costs by eliminating the need for intermediaries. Furthermore, this technology allows for real-time energy production and consumption tracking, aiding in better energy distribution and forecasting. This real-time capability also facilitates dynamic pricing based on actual demand and supply, ensuring that consumers are billed accurately (Yildizbasi 2021). Additionally, blockchain verification processes play a pivotal role in fraud detection, drastically reducing billing errors and other fraudulent activities.

The intersection of blockchain and energy also opens new avenues for investments. Initial coin offerings (ICOs) and token sales have emerged as novel methods of raising capital for innovative energy projects. Some energy-based ICOs have already showcased the potential of blockchain in energy financing. Moreover, blockchain's ability to easily verify and trade green credentials has given rise to new investment instruments, such as green bonds and renewable energy certificates. This technology also stimulates infrastructure development, particularly in the domains of smart grids and Internet of Things (IoT) devices. Market dynamics are also experiencing a seismic shift due to blockchain. Traditional centralised energy markets are giving way to decentralised models, where energy can be produced and consumed locally. This decentralisation is further bolstered by P2P energy trading, which allows consumers to sell excess energy directly. Real-time pricing mechanisms adjust to immediate market demands, benefiting consumers while ensuring grid stability (Borowski 2021). Furthermore, the introduction of blockchain is paving the way for new business models, including decentralised energy service companies that operate differently from traditional structures.

However, this transformative journey is not without challenges. Regulatory concerns loom large, with a pressing need for clear legal frameworks to accommodate and regulate blockchain growth in the energy sector. Technical limitations, particularly the scalability issues and the energy-intensive nature of some blockchain mechanisms, pose further challenges. Finally, market adoption remains a hurdle, with resistance expected from established players, and a constant need to build consumer trust and awareness. Several real-world examples have emerged in which companies have successfully integrated blockchain within their energy operations. These case studies provide invaluable information on the economic impact of technology, highlighting its potential benefits and pitfalls. In conclusion, the intertwining of blockchain and the energy sector heralds a new era of opportunities and challenges (Ahl *et al* 2022). The transformative potential of this integration is vast and promises a future where market dynamics, efficiency, and investment opportunities are redefined.

6.4.2 The regulatory landscape of blockchain's application in the energy sector across different countries and regions

The application of blockchain in the energy sector is rapidly gaining prominence. As this emergent technology continues to evolve, there is an increasing need for comprehensive regulations to guide its growth and ensure its responsible integration into contemporary energy systems. The intersection of blockchain and energy offers transformative possibilities, from enabling advanced smart contracts to reshaping the landscape of energy trading. However, these decentralised energy systems, though replete with potential, introduce challenges that regulators must address due to the novelty and intricacy of the technology. One key regulatory consideration is consumer protection. While blockchain inherently offers a degree of transparency, it simultaneously poses potential risks to consumers (Ahl *et al* 2022). There have been cases where consumers suffered fraud or losses in sectors that use blockchain, which highlights the urgent need for measures to ensure transparency and accountability, especially in the realm of P2P energy trading.

Data privacy and security are equally significant. The immutable character of the blockchain poses challenges in terms of data privacy. Regulatory frameworks, such as the General Data Protection Regulation (GDPR), have already had substantial implications on how blockchain can process and manage data, with these concerns accentuated in the energy domain. Maintaining market integrity and fostering competition are also paramount. In decentralised markets, avoiding any form of manipulation becomes crucial, which suggests the need for stringent licensing and regulatory oversight for blockchain-aided energy services (Nwaiwu 2021). Furthermore, as we stand on the brink of a new era of cross-border energy trading, there is an undeniable requirement for universally accepted technical standards that ensure compatibility and smooth functioning across diverse systems.

Regionally, the approach to blockchain in the energy sector exhibits considerable variation. In North America, the United States has federal entities such as the FERC and SEC that play a pivotal role in guiding the application of blockchain in energy. States such as New York have even carved out distinctive approaches to this paradigm. Meanwhile, Canada has provincial variations in energy regulations but relies on national energy boards for overarching guidance. Europe presents another intricate landscape. The European Union has aligned its energy strategy with the European Green Deal, providing a context for blockchain's role in energy (Borges *et al* 2022). Germany, for instance, is integrating blockchain within its Energiewende initiative, while France remains focused on nurturing decentralised energy systems. In Asia-Pacific, China's proactive promotion of blockchain coexists with its tight regulatory hold on the energy sector. At the same time, Japan, in its post-Fukushima phase, is leaning toward decentralisation and blockchain, and South Korea is innovatively testing waters via regulatory sandboxes. In Latin America, Brazil is seeing evolving governmental stances on blockchain in energy, guided by established regulatory entities. In Africa, South Africa is exploring decentralised blockchain solutions amidst ongoing energy crises, and Nigeria is harnessing its burgeoning tech scene to bring blockchain to the forefront of its energy concerns (Teng *et al* 2021). Meanwhile, GCC countries in the Middle East are transitioning from traditional oil-centric models, necessitating evolved regulatory frameworks to incorporate blockchain into their energy vision.

In examining this global landscape, it becomes evident that while some nations are trailblazers in adopting blockchain in their energy sectors, they offer valuable lessons and best practices. Others who faced challenges can provide insights into potential areas of improvement or caution. Globally, diverse case studies highlight the types of regulatory environments conducive to blockchain's successful adoption in the energy sector. However, there are also cautionary tales where potentially overbearing regulations might have stifled innovation (Kuo and Shyu 2021). In summary, as blockchain and the energy sector converge, the ongoing collective learning and regulatory evolution promise to shape a future that is sustainable, efficient and brimming with innovation.

6.4.3 The environmental impact of blockchain on energy

The modern age of technology and data management is underscored by the rise of blockchain, a decentralised ledger system. While blockchain's transformative potential spans diverse sectors, it is paramount to appreciate the environmental implications, particularly concerning energy consumption, which arises due to its operational demands. Blockchain inherently demands energy, which becomes pronounced when we delve into the mechanics of its operation. Different blockchain structures, notably consensus mechanisms, determine the energy intensity. For instance, PoW is particularly energy-hungry, requiring vast computational power to validate transactions. This contrasts with PoS and Delegated Proof-of-Stake (DPoS), which have a considerably lower energy footprint. Beyond consensus, node operations, especially full nodes responsible for maintaining the complete blockchain ledger, add to the energy tally (Ibañez and Freier 2023). The very essence of a distributed ledger, which the blockchain embodies, means that there are multiple replicas of the ledger, which compounds the environmental impact.

To grasp the magnitude of this problem, consider large-scale blockchain operations. Dedicated data centres running blockchain operations can be particularly energy intensive. Compared to traditional data centres, they can sometimes consume more energy given the complex calculations, especially in PoW systems. The carbon footprint is another concern. With substantial electricity consumption, CO2 emissions from blockchain operations can be significant. The supporting infrastructure, such as cooling systems for these data centres, compounds the energy burden. In addition, the hardware side is not benign. The servers and hardware dedicated to blockchain consume power, and can contribute to e-waste when their life cycles end. Interestingly, while blockchain itself can be an energy guzzler, its application in the energy sector offers an ironic twist. Blockchain can enable greener energy markets. It can revolutionise P2P energy trading, ensuring more efficient energy usage and reducing transmission losses. Furthermore, blockchain's indelible nature means that it can robustly track and certify renewable energy, ensuring authenticity (Abu *et al* 2023). Imagine smart grids optimised through blockchain, distributing energy efficiently and allowing dynamic pricing based on real-time data, driving both economic and environmental efficiency.

Given the environmental concerns, there is a growing focus on creating a sustainable blockchain ecosystem. Prioritising less energy-intensive consensus mechanisms, such as PoS and DPoS, can substantially reduce the energy imprint. In terms of infrastructure, embracing energy-efficient servers, considering geographic locations for natural cooling, and employing efficient cooling solutions can reduce energy consumption. There is also a clarion call to transition blockchain data centres to renewable energy. By doing so, not only will the operational carbon footprint be reduced but the entire ecosystem can also champion sustainable practices.

Additionally, initiatives that offset carbon can further neutralise the environmental impact. In real-world contexts, blockchain is already being harnessed for environmental good. Supply chains take advantage of blockchain to ensure ethical and green sourcing of products, bolstered by the transparency and immutability of the ledger. Certain environmental monitoring projects are based on blockchains to maintain data integrity, ensuring authentic and tamper-proof records. Moreover, renewable energy projects find value in blockchain's transparent nature, optimising energy distribution and usage. In conclusion, while blockchain does pose environmental challenges, primarily due to its energy-intensive operations, its potential to drive environmental efficiency, especially in the energy sector, is undeniable (Lal and You 2023). Addressing the energy concerns, combined with its green applications, can position blockchain as a technological marvel and an environmental ally.

6.4.4 The intersection of Blockchain with other technologies

In the evolving landscape of the energy sector, the blockchain is not an isolated player. Instead, it is at a crossroads, intertwining with other cutting-edge technologies such as Artificial Intelligence (AI), IoT, and big data. Recognising how these technologies interact is pivotal to unlocking unprecedented efficiencies and insights. At the juncture of blockchain and AI lies a realm of possibilities. With its prowess in energy forecasting, optimisation, and anomaly detection, AI can feed off the transparent and immutable data provided by blockchain. This synergy enables AI algorithms to derive accurate predictions and decisions, enriching the decision-making process. In addition, smart contracts on blockchain platforms can be augmented with AI capabilities, making them even more dynamic and adaptable (Rejeb *et al* 2023). For example, there are ongoing projects where AI-driven energy consumption predictions are recorded on blockchain, ensuring transparent and accountable energy utilisation.

The impact in the energy sector is undeniable. With smart meters, energy consumption monitoring tools, and the broader smart grid infrastructure, IoT devices are now widespread. However, with the proliferation of these devices comes the concern of security and data integrity. Enter the blockchain. By offering a decentralised platform, blockchain can verify the identity of IoT devices and guarantee the authenticity of the data they generate. A transformative application emerges in the form of P2P energy trading. Using data from IoT devices, the blockchain can facilitate decentralised energy markets, where consumers can trade excess energy without intermediaries. Many projects around the world are experimenting with such solutions, leading to democratised and efficient energy distribution. Given the vast amounts of information generated in the energy sector, big data holds the promise of game-changing insights. However, the veracity of these data is crucial. Blockchain ensures that massive data sets are genuine, facilitating transparent and tamper-proof data sharing. Once stored on a blockchain, data analytics tools can sift through blockchain transactions and patterns to forecast energy market dynamics or pinpoint inefficiencies (Kong *et al* 2022). Several energy companies have started integrating their data repositories with blockchain to achieve these results.

Imagine a future where AI, IoT, blockchain, and big data converge, creating an energy ecosystem that is greater than the sum of its parts. Such unified systems can facilitate the predictive maintenance of infrastructure using AI algorithms that feed off IoT-generated data stored on a blockchain. Real-time optimisation of energy grids becomes achievable, and decentralised trading based on vast data insights becomes the norm. However, the marriage of these technologies is not without challenges. Integrating them requires the inherent complexities to be addressed, ensuring smooth interoperability and managing the huge influx of data. However, with continuous advancements and collaborations, these hurdles are not insurmountable. In essence, as the blockchain forges alliances with AI, IoT, and big data in the energy sector, a new paradigm of efficiency, transparency, and democratisation emerges (Lyu and Liu 2021). This interconnected future promises to reshape the energy landscape, drive sustainability, and empower consumers.

6.4.5 Evaluating blockchain and energy nexus challenges

Blockchain's promise in the energy sector is immense, from decentralised energy trading to transparent carbon credit markets. However, this revolutionary interplay is not without its challenges, ranging from technical hurdles to social concerns. On the technical front, scalability remains a significant impediment. Since energy systems generate vast data streams, current blockchain structures often struggle to keep up with them. While research is underway to develop more scalable solutions, breakthroughs are yet to become mainstream. Closely linked is the interoperability issue, with the integration of various blockchain platforms and traditional energy systems presenting considerable challenges. Developers are now keenly looking into cross-chain solutions to ensure seamless interactions. Moreover, latency in blockchain systems is a pressing concern, especially for applications demanding real-time decisions, such as energy trading (Bekrar *et al* 2021). Efforts to increase transaction speeds are vital to unlocking blockchain's full potential in the sector.

From a regulatory perspective, consumer protection is an important topic. Blockchain must ensure secure energy transactions while balancing the need for transparency with user data protection. Moreover, since the energy sector has long been rooted in centralised infrastructures, integrating decentralised blockchain systems poses significant challenges. Security concerns are also paramount. From potential 51% attacks to vulnerabilities in smart contracts, the sector needs rigorous solutions to ensure data integrity and security. Economically, resistance from entrenched energy market players threatens blockchain adoption. Coupled with this is the need for a robust understanding of the cost implications of blockchain integrations. Assessing the return on investment and the long-term viability of such projects is crucial (Wongthongtham *et al* 2021). In addition, the creation and management of energy tokens or similar assets present a unique set of challenges, necessitating clear guidelines and systems.

Public trust in decentralised solutions remains a significant barrier at the societal level. Building this trust requires broad stakeholder engagement, bringing together producers, consumers, regulators, and innovators. Addressing the prevalent knowledge gap about blockchain in the energy sector through education is crucial. Environmental concerns also surface, especially considering the energy-intensive operations of certain blockchain types, such as those employing PoW mechanisms. Assessing and minimising the environmental footprint of blockchain within the energy sector is vital to its sustainable adoption. Looking ahead, a combination of technological innovations, intelligent policy making, and collaborative efforts could be the key to navigating these challenges. Technological advances, especially in scalability and security, can pave the way for a more extensive adoption. Regulators have a crucial role in crafting policies that strike a balance between fostering innovation and ensuring robust consumer protection (Wongthongtham *et al* 2021). Finally, collaborative endeavours, particularly consortiums and industry groups, can pool resources and knowledge, offering comprehensive solutions to the challenges at the blockchain-energy nexus.

6.5 Summary

The convergence of blockchain and the energy sector presents a transformative potential that could redefine the way in which energy systems operate, offering decentralisation, transparency, and enhanced efficiency. Nevertheless, this nexus is not without its complexities, from technical and regulatory challenges to social and environmental considerations. For blockchain to achieve its full potential within the energy sector, these challenges must be addressed methodically. Embracing technological advancements, creating informed policies and fostering collaborative platforms can be instrumental in this journey. As the energy landscape continues to evolve, the interplay of blockchain will undeniably play a pivotal role in shaping the future, making proactive addressing of its associated challenges not only necessary but imperative for a sustainable and inclusive energy future.

References

Abu M, Amaya M and Marston L 2023 The water and carbon footprint of cryptocurrencies and conventional currencies *J. Clean. Prod.* **411** 137268–8

Ahl A, Goto M, Yarime M, Tanaka K and Sagawa D 2022 Challenges and opportunities of blockchain energy applications: interrelatedness among technological, economic, social, environmental, and institutional dimensions *Renew. Sustain. Energy Rev.* **166** 112623

Bekrar A, Ait El Cadi A, Todosijevic R and Sarkis J 2021 Digitalizing the closing-of-the-loop for supply chains: a transportation and blockchain perspective *Sustainability* **13** 2895

Bhutta M N M, Khwaja A A, Nadeem A, Ahmad H F, Khan M K, Hanif M, Song H, Alshamari M and Cao Y 2021 A survey on blockchain technology: evolution, architecture and security *IEEE Access* **9** 61048–73

Borges C E, Kapassa E, Touloupou M, Legarda Macón J and Casado-Mansilla D 2022 Blockchain application in P2P energy markets: social and legal aspects *Connect. Sci.* **34** 1066–88

Borowski P F 2021 Digitization, digital twins, blockchain, and Industry 4.0 as elements of management process in enterprises in the energy sector *Energies* **14** 1885

Diffie W and Hellman M E 1976 New Directions in Cryptography *IEEE Transactions on Information Theory* **22** 644–54

Ferrón-Vílchez V, Valero-Gil J and Suárez-Perales I 2020 How does greenwashing influence managers' decision-making? An experimental approach under stakeholder view *Corp. Soc. Responsib. Environ. Manag.* **28** 860–80

Foti M and Vavalis M 2021 What blockchain can do for power grids? *Blockchain: Res. Appl.* **2** 100008

George K and Patatoukas P N 2020 The blockchain evolution and revolution of accounting *Inform. Effi. Decis. Mak.* 157–72

Gorkhali A and Chowdhury R 2021 Blockchain and the evolving financial market: a literature review *J. Ind. Integr. Manag.* **7** 47–81

Gawusu S, Zhang X, Ahmed A, Jamatutu S A, Miensah E D, Amadu A A and Osei F A J 2022 Renewable energy sources from the perspective of blockchain integration: from theory to application *Sustain. Energy Technol. Assess.* **52** 102108

Habib G, Sharma S, Ibrahim S, Ahmad I, Qureshi S and Ishfaq M 2022 Blockchain technology: benefits, challenges, applications, and integration of blockchain technology with cloud computing *Future Internet* **14** 341

Ibañez J I and Freier A 2023 Bitcoin's carbon footprint revisited: proof of work mining for renewable energy expansion *Challenges* **14** 35

Juszczyk O and Shahzad K 2022 Blockchain technology for renewable energy: principles, applications and prospects *Energies* **15** 4603

Kong X, Wu Y, Wang H and Xia F 2022 Edge computing for internet of everything: a survey *IEEE Internet Things J.* **9** 23472–85

Kuo C-C and Shyu J Z 2021 A cross-national comparative policy analysis of the blockchain technology between the USA and China *Sustainability* **13** 6893

Krithika L B 2022 Survey on the applications of blockchain in agriculture *Agriculture* **12** 1333

Lal A and You F 2023 Climate concerns and the future of nonfungible tokens: leveraging environmental benefits of the Ethereum Merge *Proceedings of the National Academy of Sciences* **120** e2303109120

Lyu W and Liu J 2021 Artificial intelligence and emerging digital technologies in the energy sector *Appl. Energy* **303** 117615

Maka A O M and Alabid J M 2022 Solar energy technology and its roles in sustainable development *Clean Energy* **6** 476–83

Nakamoto S 2008 Bitcoin: *A peer-to-peer electronic cash system. Decentralized business review*

Nwaiwu F 2021 Digitalisation and sustainable energy transitions in Africa: assessing the impact of policy and regulatory environments on the energy sector in Nigeria and South Africa *Energy, Sustain. Soc.* **11** 1–16

Parmentola A, Petrillo A, Tutore I and De Felice F 2021 Is blockchain able to enhance environmental sustainability? A systematic review and research agenda from the perspective of Sustainable Development Goals (SDGs) *Bus. Strategy Environ.* **31** 194–217

Reddy P T S and Rao S V A 2023 Web3: unleashing the power of decentralization and blockchain technology to transform the digital landscape *Turk. J. Comput. Math. Educ. (TURCOMAT)* **14** 1036–43 https://turcomat.org/index.php/turkbilmat/article/view/13931 (Accessed 21 August 2023)

Rejeb A, Rejeb K, Appolloni A, Jagtap S, Iranmanesh M, Alghamdi S, Alhasawi Y and Kayikci Y 2023 Unleashing the power of internet of things and blockchain: a comprehensive analysis and future directions *Internet Things Cyber-Phys. Syst.*

Teng F, Zhang Q, Wang G, Liu J and Li H 2021 A comprehensive review of energy blockchain: application scenarios and development trends *Int. J. Energy Res.* **45** 17515–31

Valdivia A D and Balcell M P 2022 Connecting the grids: a review of blockchain governance in distributed energy transitions *Energy Res. Soc. Sci.* **84** 102383

Wang Q, Li R and Zhan L 2021 Blockchain technology in the energy sector: from basic research to real world applications *Comput. Sci. Rev.* **39** 100362

Wongthongtham P, Marrable D, Abu-Salih B, Liu X and Morrison G 2021 Blockchain-enabled peer-to-peer energy trading *Comput. Electr. Eng.* **94** 107299

Yildizbasi A 2021 Blockchain and renewable energy: integration challenges in circular economy era *Renew. Energy* **176** 183–97

IOP Publishing

Blockchain-based Peer-to-Peer Transactions in Energy Systems

Nnamdi Nwulu and Uyikumhe Damisa

Chapter 7

Blockchain-based peer-to-peer energy trading through a double auction mechanism

The increased adoption of renewable energy generation by both households and utilities requires a change in the operational models of the modern power grid. With the help of peer-to-peer (P2P) systems underpinned by blockchain, households can sell energy generated to peers in the power grid. Smart contract technology is crucial to the implementation of these systems. In this chapter, a smart contract, which was originally developed for P2P energy trading in a previously published work, is tested with data depicting different auction market scenarios. In addition to the tests performed, a bug in the smart contract is fixed.

7.1 Introduction and literature review

The massive adoption of renewable energy sources has led to the decentralisation of power generation in many power grids around the world, which has empowered consumers of electricity to contribute to the generation mix. The modern grid is replete with distributed energy resources such as generators, batteries, electrical vehicles, and so on, some of which contribute to the stability of the grid. In particular, the output of renewable energy generators such as solar and wind tends to fluctuate, hence there will be periods when they supply excess or deficit energy, in which case the consumer may choose to sell or buy energy from their neighbours. Electricity consumers who also produce electrical energy and can sometimes sell to the grid are known as prosumers. The emergence of new technological innovations, increased participation, and the rise in penetration of renewable energy sources at the distributed and bulk generation levels are driving a paradigm shift in the electricity industry (Rahimi *et al* 2016). This shift has seen household consumers become active players in the electricity market through their participation in energy trade transactions. Centralised systems can be used to facilitate such transactions;

doi:10.1088/978-0-7503-6295-5ch7

however, they are plagued by a number of downsides. The traditional use of centralised schemes to facilitate these transactions leads to management inefficiencies and increased transaction costs (Han *et al* 2020). They are also prone to hacking and censorship. Privacy is another concern in centralised systems. A feasible alternative is a P2P system, which enables transactions between participants to be performed without the involvement of third parties. In the context of prosumers connected to a power grid, a P2P system facilitates energy trades between electricity consumers and neighbouring prosumers. This creates virtual local electricity markets, which offer benefits. Load balancing in a renewable energy-assisted grid with variable demand can be achieved securely and economically by P2P energy trading because it involves matching energy production with consumption locally (Li *et al* 2020). The Piclo platform, Vandebron platform, and Sonnen community are real-world P2P energy trading schemes. However, the centralised database architectures that are employed by some of these existing P2P energy trading systems can cause problems for the privacy of consumer data and suffer from clumsy transactions (Han *et al* 2020). In addition, issues around scalability arise in centralised schemes in the case of a strong increase in participating prosumers (Siano *et al* 2019).

Blockchain technology possesses inherent characteristics that make it suitable for powering P2P solutions in the energy industry. In a blockchain-based network, the transactions carried out by network participants are recorded in an electronic ledger that is distributed among key participants. This ledger eliminates the need for third-party services in the processing of digital transactions (Monroe *et al* 2020). Blockchains employ consensus mechanisms that offer optimal solutions and high computing power when incorporated into energy trading schemes (Wang *et al* 2019). The literature contains studies on blockchain applications from many different perspectives. A blockchain-based industrial Internet of Things (IoT)-integrated energy trading system, which protects the information of electricity consumers and ensures electricity market demand and supply balance, is proposed in Guan *et al* (2020). In Hu and Li (2021), a suitable blockchain-based framework for P2P energy transactions was presented for typical industrial IoT transaction scenarios. In a P2P network based on blockchain, transactions can be public while the security and privacy of network participants are preserved (Abdella and Shuaib 2018). A blockchain-based and privacy-preserving energy trading scheme is proposed in Khorasany *et al* (2021), where network participants are encouraged to trade with closer peers through a pricing mechanism that factors in the electrical distance between users. In Lin (2019), a Hyperledger Fabric-based, hour-ahead trading network is presented. Different auction mechanisms are also compared. The authors of AlSkaif *et al* (2021) proposed new methods to determine bilateral trading inclinations of residences in a local energy market. The Hyperledger Fabric platform was adopted, and studies showed a reduction in the cost of energy procurement. A general framework based on blockchain technology for P2P trading in retail electricity markets is presented in Han *et al* (2020), which used an Ethereum private chain and shows that the proposed platform effectively and efficiently facilitates transactions among several users. Kwak and Lee (2021) designed an Ethereum-based platform that enabled distant users to engage in P2P energy transfers. A hybrid Ethereum-based P2P energy trading marketplace is implemented in Khalid *et al* (2020). The proposed scheme resulted in a reduction in the energy costs. Esmat *et al* (2021) proposed a blockchain-based double-layered energy trading platform in which automated settlements and security are offered by the blockchain layer, while the auction and clearing scheme are featured on the market layer. Hua *et al* (2020) deployed blockchain-based smart contracts beyond energy trading and incorporated carbon allowance trading. They proposed an Ethereum-based P2P framework for the trading of energy and carbon allowances. Meanwhile, van Leeuwen *et al* (2020) demonstrated the use of blockchain for energy management. Li *et al* (2019) proposed an integrated platform for energy management and a bilateral trade scheme which incorporated smart contracts based on blockchain to manage a distributed energy environment. In Foti *et al* (2021), the blockchain-based consensus algorithm has been explored to offer a decentralised solution approach to the optimal power flow problem. Smart contract technology opens blockchain technology to a wider range of application areas. With smart contracts based on blockchain technology, credible transactions can be facilitated without third-party participation (Han *et al* 2020). Fairness and security are improved in energy trading because smart contracts can enforce trading and payment guidelines without human involvement (Han *et al* 2020). A smart contract-automated bidding procedure for smart city transactions regarding energy supply and demand was presented in Khattak *et al* (2020). Blockchain-based smart contract technology was incorporated in the decentralised microgrid electricity market scheme presented in Sabounchi and Wei (2018). Incorporating blockchain technology, Esmat *et al* (2021) partitioned a grid into microgrids and proposed a multilevel trading scheme. The excess or deficit energy obtained after trading at the intra-microgrid level is traded between microgrids. This chapter builds on the work presented in Damisa *et al* (2022).

7.2 Smart contract-based peer-to-peer energy trading architecture and algorithm

Figure 7.1 depicts a high-level representation of a P2P energy trading scheme. A more detailed representation is given in figure 7.2. Energy producers and willing buyers send their offers and bids, respectively, to a smart contract. The contract matches energy producers and buyers based on its pre-programmed auction procedure. While producers send only data to the smart contract, buyers send funds as well as data. The smart contract holds these funds until the energy has been delivered by the sellers; after which payment is made to the sellers. Intending buyers who, as a requirement, submitted bids along with funds and are unsuccessful after an auction are refunded by the smart contract. In implementing this scheme, meters may be required at the participants' premises to communicate with the smart contract.

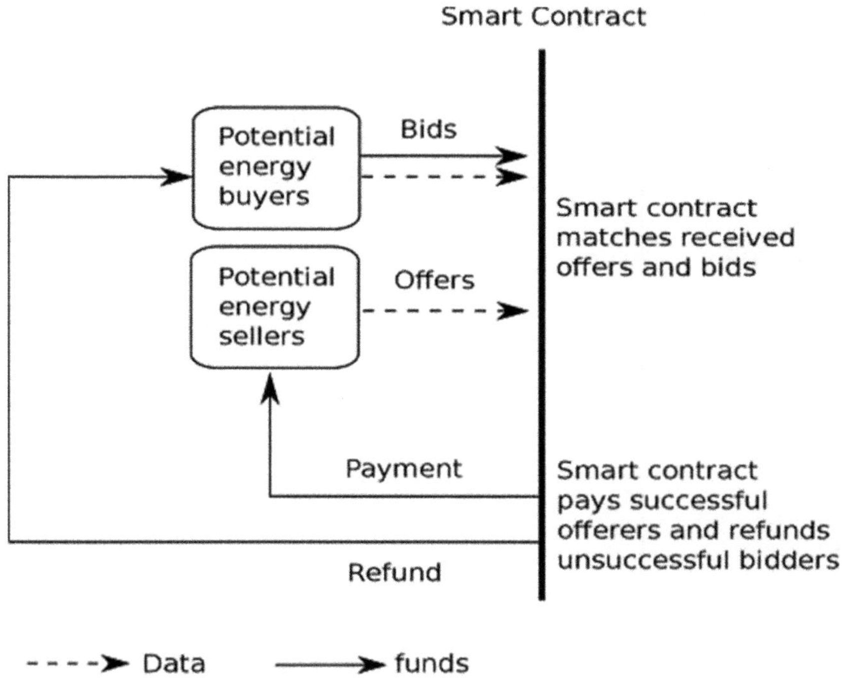

Figure 7.1. Block diagram of a P2P energy trading scheme. Reproduced from Damisa *et al* (2022). CC BY 4.0.

7.3 Smart contract functions for peer-to-peer energy trading scheme

The following functions are contained in the smart contract-based P2P energy trading scheme.

Constructor: This function gets executed at the point of contract deployment. In the case at hand, the address of the contract deployer is assigned to a variable 'owner'.

setRate: This allows the dollar equivalent of an ether to be set. The function restricts the privilege of setting this value to the 'owner'.

CheckEtherEquiv: With this function, an intending bidder can view the cost in ethers of the amount of energy being demanded.

offers: Energy sellers submit their energy amount and price offers to this function.

verifiedOffers: The offers received in the foregoing function are ordered in ascending order based on price and returned to this function. The function checks that the returned data matches the original data received by offers to ensure data integrity.

bids: This function receives energy amounts, price bids, and funds sufficient to meet the required energy.

verifiedBids: The bids received in the preceding function are ordered in a descending order based on price and returned to this function. The function

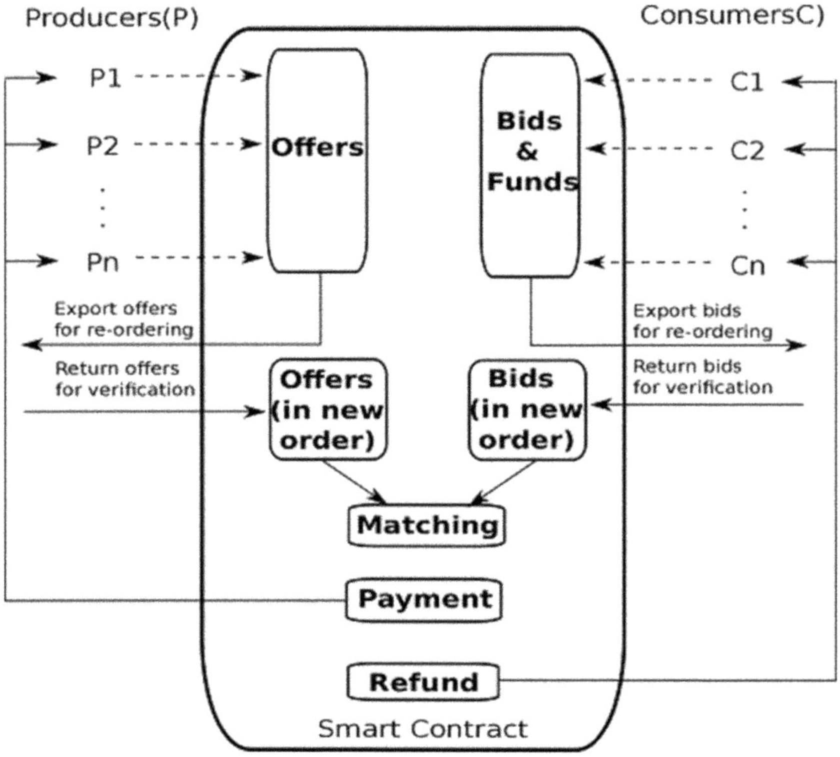

Figure 7.2. Detailed illustration of a P2P energy trading scheme Reproduced from Damisa *et al* (2022). CC BY 4.0.

checks that the returned data matches the original data received by bids to ensure data integrity.

matching: This is a crucial function because it determines the winners of an auction round. It uses a pre-programmed auction procedure to match offerers and bidders, and determines the price at which each pair transacts. Figure 7.3 shows the smart contract flowchart of the matching function.

paySellers: This function is called by the owner to instruct the smart contract to make payments to successful energy sellers.

refundBuyers: The owner also calls this function to refund bidders who were unsuccessful during an auction round.

The P2P trading smart contract scheme proposed in Damisa *et al* (2022) is vulnerable at the point where ordered offer data are being returned to the smart contract. While it ensures that the returned data matches the original data submitted by offerers, no measure is put in place to guarantee that the data being returned are in the right order. Hence, an attacker can tamper with the order of the offers outside the smart contract to favour certain offerers, giving them a higher priority than other

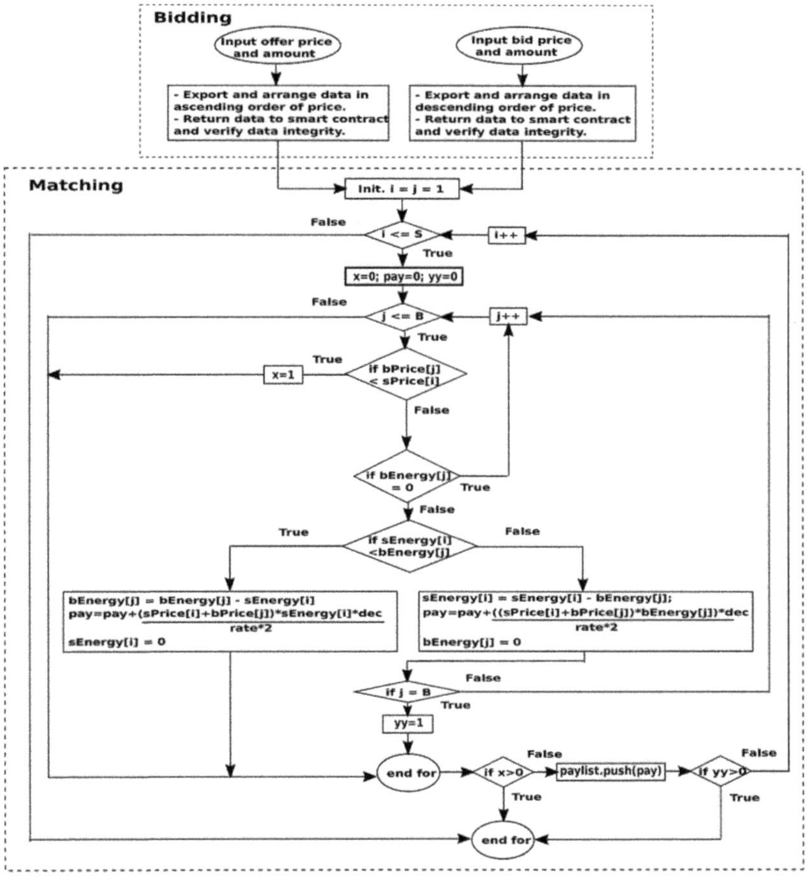

Figure 7.3. Matching function flowchart of a P2P energy trading scheme. Reproduced from Damisa *et al* (2022). CC BY 4.0.

cheaper offers. This chapter introduces a measure to check that the return offer data is rightly ordered.

7.4 Case study and discussion

In this study, we investigate the smart contract using three market scenarios: a base scenario, price scenarios, and quantity scenarios. While the base scenario is obtained from the literature, the price and quantity scenarios are hypothetically constructed to represent extreme energy market conditions, which tests the ability of the smart contract to handle such cases.

7.4.1 Base scenario

The base scenario is obtained from Fang *et al* (2012). See Table 2 of Damisa *et al* (2022) for the base scenario data. Potential energy sellers and buyers submit their offers and bids, respectively, to the smart contract. The offers are arranged from the

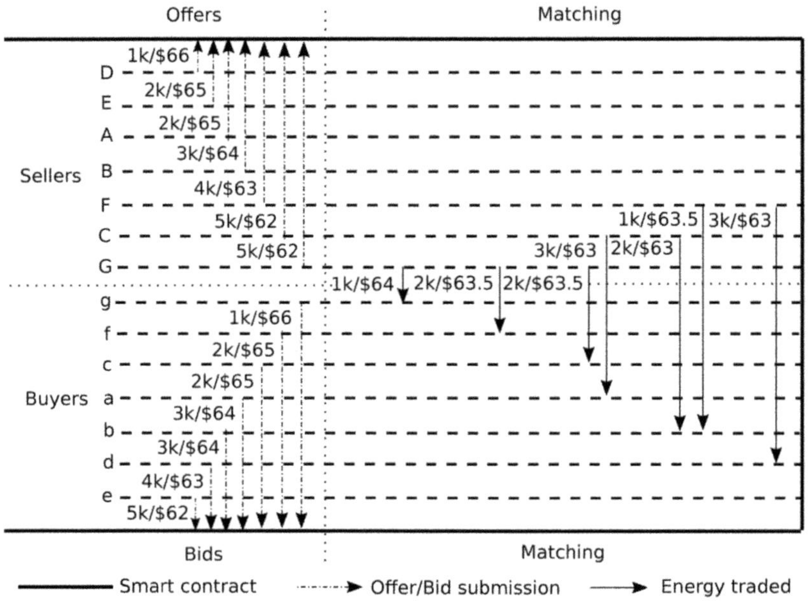

Figure 7.4. Smart contract execution for matching in a P2P energy trading scheme. Reproduced from Damisa *et al* (2022). CC BY 4.0.

lowest offer price to the highest, while the bids are arranged from the highest to the lowest price. Matching commences from the lowest offerer and the highest bidder, and the clearing price for each matched pair is the midpoint between the offer and bid prices. If the bid price falls below the offer price at any point as matching progresses, then matching is terminated and the auction round is closed. Unsuccessful bidders get a refund of their committed funds. Figures 7.4 and 7.5 show the smart contract's execution of the auction procedure, payments, and refunds.'

7.4.2 Price scenarios

The first case in the price scenarios investigates the response of the smart contract to cases where all of the bid prices are higher than the offers, i.e., the lowest bid price is more than the highest offer price (table 7.1). In the second case, the lowest offer price exceeds the highest offer price (table 7.2). 1 Ether is taken to be equivalent to $3000.

As expected, when the offers are all higher than the available bids, then no buyer/seller matches in the auction round. This results in a loss of the transaction fees incurred by all the parties involved in the scheme. Generally, the smart contract administration fees include deploying the contract, setting the dollar equivalent of an Ether, verifying ordered offers/bids, matching buyers and sellers, and paying and refunding sellers and unsuccessful buyers, respectively. A potential seller pays a transaction fee to submit an offer to the smart contract, while a potential buyer is charged a transaction fee to submit a bid. Some or all of these fees can be lost in an auction round, such as the present one being considered. Mechanisms can be put in

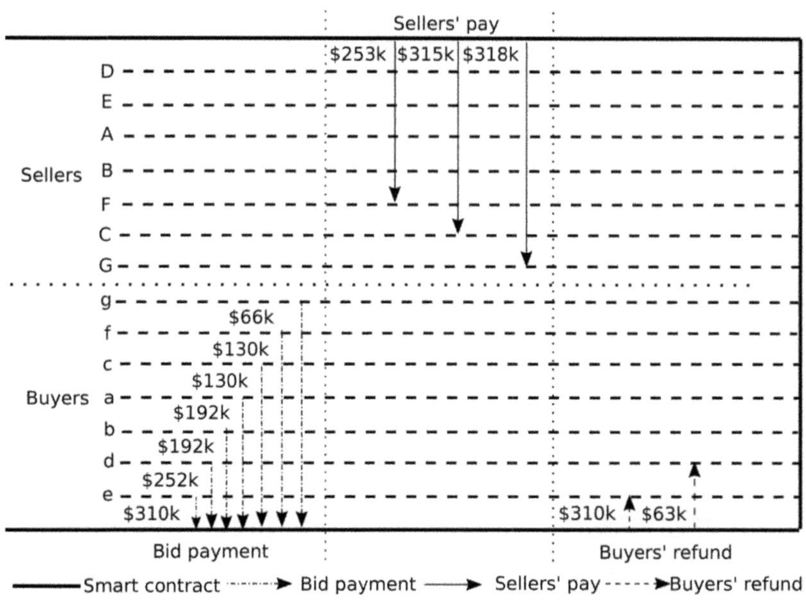

Figure 7.5. Smart contract execution of payments and refunds in a P2P energy trading scheme. Reproduced from Damisa *et al* (2022). CC BY 4.0.

Table 7.1. Sellers/buyers' offer/bid amount and prices (offers<bids).

Seller	offer ($ MW h^{-1})	Quantity (MWh)	Buyer	Bid ($ MW h^{-1})	Quantity (MWh)
A	63	2000	a	67	3000
B	64	3000	b	65	3000
C	62	5000	c	65	2000
D	60	1000	d	67	4100
E	61	2000	e	66	3000

Table 7.2. Sellers/buyers' offer/bid amount and prices (offers>bids).

Seller	Offer ($ MW h^{-1})	Quantity (MWh)	Buyer	Bid ($ MW h^{-1})	Quantity (MWh)
A	66	2000	a	61	3000
B	67	3000	b	64	3000
C	65	5000	c	61	2000
D	66	1000	d	63	4000
E	65	2000	e	62	2000

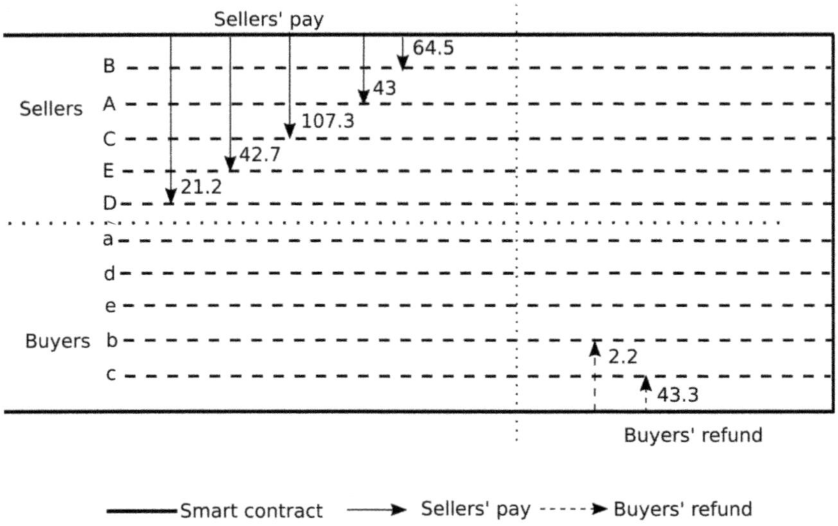

Figure 7.6. A smart contract's payments in ethers.

Table 7.3. Sellers/buyers' offer/bid amount and prices (demand > supply).

Seller	Offer ($ MW h^{-1})	Quantity (MWh)	Buyer	Bid ($ MW h^{-1})	Quantity (MWh)
A	65	2000	a	64	3000
B	64	3000	b	64	3000
C	62	500	c	65	2000
D	66	1000	d	63	4000
E	65	2000	e	62	2000

place to narrow the spread between offers and bids to avoid such a scenario. Larger clusters of potential sellers and buyers may also reduce the chances of realising this scenario. Figure 7.6 shows the payments made in Ethers by the smart contract to the appropriate parties.

7.4.3 Quantity scenarios

The first scenario in the quantity scenarios is one in which the total amount of energy demanded is greater than is available (table 7.3). The second assumes that the energy supply exceeds the demand (table 7.4).

The quantity scenarios investigated affirm that price is the major consideration when matching sellers and buyers. Regardless of the quantities available or demanded, offer and bid prices determine how matching is performed. Matching is impossible if a bid price is lower than an offer price. Consequently, there could be both energy supply and demand that are not matched due to price. This further

Table 7.4. Sellers/buyers' offer/bid amount and prices (demand < supply).

Seller	Offer ($ MW h^{-1})	Quantity (MWh)	Buyer	Bid ($ MW h^{-1})	Quantity (MWh)
A	65	3000	a	64	2000
B	64	3000	b	64	3000
C	62	2000	c	65	500
D	66	4000	d	63	1000
E	65	5000	e	62	2000

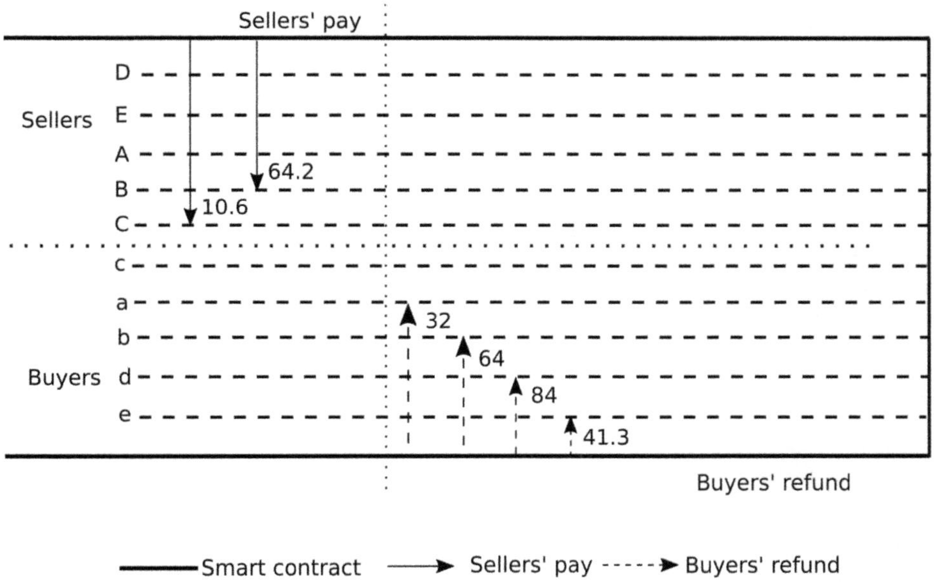

Figure 7.7. A smart contract's payments in ethers.

strengthens the need for mechanisms that narrow the spread between offers and bids. Figures 7.7 and 7.8 show the payments made in Ethers by the smart contract to the appropriate parties.

7.5 Summary

A successfully implemented blockchain-based P2P energy trading system for renewable energy dominated power grids offers significant benefits. It brings transparency and trustlessness to P2P energy transactions, which in turn drives further renewable energy adoption. The work presented in this chapter is only a step towards such a system. The smart contract has been tested and verified to accurately perform energy auctions and effect payments. However, this test was done by manually interacting with the contract. It would need to be enhanced in a practical setting for more

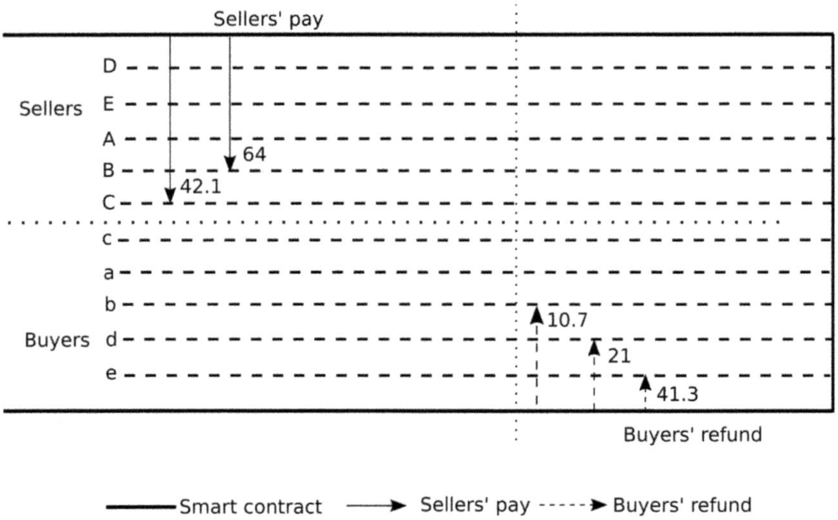

Figure 7.8. A smart contract's payments in ethers.

autonomy. For instance, the dollar equivalent of an ether can be sent to the smart contract via an independent Oracle. Payment to the seller can also be automated upon confirmation that the agreed amount of energy was delivered.

References

Abdella J and Shuaib K 2018 Peer to peer distributed energy trading in smart grids: a survey *Energies* **11** 1560

AlSkaif T, Crespo-Vazquez J L, Sekuloski M, van Leeuwen G and Catalão J P S 2021 Blockchain-based fully peer-to-peer energy trading strategies for residential energy systems *IEEE Trans. Ind. Inf.* **18** 231–41

Damisa U, Nwulu N I and Siano P 2022 Towards blockchain-based energy trading: a smart contract implementation of energy double auction and spinning reserve trading *Energies* **15** 4084

Esmat A, de Vos M, Ghiassi-Farrokhfal Y, Palensky P and Epema D 2021 A novel decentralized platform for peer-to-peer energy trading market with blockchain technology *Appl. Energy* **282** 116123

Fang D, Wu J and Tang D 2012 A double auction model for competitive generators and large consumers considering power transmission cost *Int. J. Electr. Power Energy Syst.* **43** 880–8

Foti M, Mavromatis C and Vavalis M 2021 Decentralized blockchain-based consensus for optimal power flow solutions *Appl. Energy* **283** 116100

Guan Z, Lu X, Wang N, Wu J, Du X and Guizani M 2020 Towards secure and efficient energy trading in IIoT-enabled energy internet: a blockchain approach *Future Gener. Comput. Syst.* **110** 686–95

Han D, Zhang C, Ping J and Yan Z 2020 Smart contract architecture for decentralized energy trading and management based on blockchains *Energy* **199** 117417

Hu W and Li H 2021 A blockchain-based secure transaction model for distributed energy in Industrial Internet of Things *Alexandria Eng. J.* **60** 491–500

Hua W, Jiang J, Sun H and Wu J 2020 A blockchain based peer-to-peer trading framework integrating energy and carbon markets *Appl. Energy* **279** 115539

Khalid R, Javaid N, Almogren A, Javed M U, Javaid S and Zuair M 2020 A blockchain-based load balancing in decentralized hybrid P2P energy trading market in smart grid *IEEE Access* **8** 47047–62

Khattak H A, Tehreem K, Almogren A, Ameer Z, Din I U and Adnan M 2020 Dynamic pricing in industrial internet of things: blockchain application for energy management in smart cities *J. Inform. Secur. Appl.* **55** 102615

Khorasany M, Dorri A, Razzaghi R and Jurdak R 2021 Lightweight blockchain framework for location-aware peer-to-peer energy trading *Int. J. Electr. Power Energy Syst.* **127** 106610

Kwak S and Lee J 2021 Implementation of blockchain based P2P energy trading platform *2021 Int. Conf. on Information Networking (ICOIN)* pp 5–7

Li J, Ye Y and Strbac G 2020 Stabilizing peer-to-peer energy trading in prosumer coalition through computational efficient pricing *Electr. Power Syst. Res.* **189** 106764

Li Y, Yang W, He P, Chen C and Wang X 2019 Design and management of a distributed hybrid energy system through smart contract and blockchain *Appl. Energy* **248** 390–405

Lin J 2019 *Analysis of Blockchain-Based Smart Contracts for Peer-to-Peer Solar Electricity Transactive Markets* (Virginia Tech.)

Monroe J G, Hansen P, Sorell M and Berglund E Z 2020 Agent-based model of a blockchain enabled peer-to-peer energy market: application for a neighborhood trial in Perth, Australia *Smart Cities* **3** 1072–99

Rahimi F, Ipakchi A and Fletcher F 2016 The changing electrical landscape: end-to-end power system operation under the transactive energy paradigm *IEEE Power Energ. Mag.* **14** 52–62

Sabounchi M and Wei J 2018 A decentralized P2P electricity market model for microgrids *2018 IEEE Power and Energy Society General Meeting (PESGM)* pp 1–5

Siano P, De Marco G, Rolán A and Loia V 2019 A survey and evaluation of the potentials of distributed ledger technology for peer-to-peer transactive energy exchanges in local energy markets *IEEE Syst. J.* **13** 3454–66

van Leeuwen G, AlSkaif T, Gibescu M and van Sark W 2020 An integrated blockchain-based energy management platform with bilateral trading for microgrid communities *Appl. Energy* **263** 114613

Wang N, Zhou X, Lu X, Guan Z, Wu L, Du X and Guizani M 2019 When energy trading meets blockchain in electrical power system: the state of the art *Appl. Sci.* **9** 1561

Nnamdi Nwulu and Uyikumhe Damisa

Chapter 8

Blockchain-facilitated peer-to-peer energy storage capacity auctioning

Space constraints and cost are crucial considerations for the large-scale energy storage users. To reduce the impact of these issues, multiple smaller-scale storage units can be pooled for large-scale usage and to ensure that the storage capacity contributors are duly compensated. In this chapter, a smart contract that was originally adopted in the literature for an auction-based storage unit sharing is tested with different storage auction market scenarios. The results obtained from these scenarios determine the accuracy of the smart contract logic for both the auction procedure and payments. Furthermore, the security of the system is enhanced by fixing a bug in the original contract.

8.1 Introduction and review of the literature

The intermittent nature of renewable energy sources, which have flooded modern power grids, requires more sophisticated approaches to energy management. Energy storage devices have proved to be effective in supporting renewable energy-assisted power grids. Due to their fast response times, storage can provide grid stability under power mismatch conditions (Chakraborty *et al* 2018). In energy systems, including smart grids, energy storage facilities play an important role in providing grid efficiency and resilience (Dai *et al* 2021). Residential-scale batteries can be pooled to provide grid ancillary services. The optimal management of residential storage units can reduce the investment cost, as well as the energy cost (Wang *et al* 2013). Furthermore, these units can also be used to reduce the carbon footprint of the grid (Tushar *et al* 2016). The success of some energy sharing and trading systems that have emerged in recent times is partly responsible for the interest of researchers in investigating energy storage-sharing models (Dai *et al* 2021). Wang *et al* (2013) proposed the sharing of energy storage ownership among network operators and

doi:10.1088/978-0-7503-6295-5ch8

residential consumers. An algorithm for the allocation of energy storage capacity to residential units in a smart grid was proposed by Xie *et al* (2019) to reduce their electricity bills. Zhao *et al* (2019) proposed an energy storage-sharing system that allows a single energy storage unit to be virtually owned and used by multiple entities. The allocation of various products such as storage capacity, energy and (dis) charging power from energy storage is achieved using a distributed auction mechanism in Zhong *et al* (2020). Luo *et al* (2020) proposed a smart contract-based scheme to facilitate energy storage sharing in a joint market comprising entities such as renewable energy plants, energy consumers, and thermal plants. Nunna *et al* (2020) proposed an agent-based trading framework that enables energy storage systems to offer services related to grid management. Rahbar *et al* (2016) investigated the sharing of an energy storage system in a power network comprising owners of renewable energy sources. The results of their investigation show that it is considerably more profitable to share storage resources than to own them individually. Sim *et al* (2021) investigated a technique that enables peer-to-peer (P2P) sharing of energy storage capacity among users. Saini *et al* (2020) presented a cloud energy storage scheme that allows the exploitation of distributed energy storage facilities in microgrids. Wang *et al* (2020) proposed a management model for shared energy storage facilities founded on proxy signatures in a blockchain environment.

Although energy storage sharing has proved beneficial, the centralised systems that are employed to facilitate some of these sharing schemes have inherent downsides. Centralised systems are typically plagued with data privacy issues and transaction clumsiness (Han *et al* 2020). Furthermore, their use in energy transactions can potentially result in high transaction costs and management inefficiencies (Han *et al* 2020). Unlike centralised systems, which are vulnerable to cyber attacks, blockchain enabled decentralised systems can still come to a unified agreement on the state of the system despite being under attack (Münsing *et al* 2017). Blockchain technology is also capable of powering a transparent, secure, and efficient P2P system. In a P2P network based on blockchain, transactions can be public while the security and privacy of network participants are preserved (Abdella and Shuaib 2018). Energy trading schemes can benefit from the high computing power and low-cost benefits of blockchains (Wang *et al* 2019). Blockchain is also applicable in other sectors of the energy industry. For example, van Leeuwen *et al* (2020) have demonstrated energy management using an integrated platform for energy management and a bilateral trade scheme. Li *et al* (2019) used smart contract technology to manage energy in a distributed environment. Meanwhile, the optimal power flow problem is solved in Foti *et al* (2021) using a decentralised, blockchain-based solution. Energy trading has also been significantly impacted by blockchain technology (Guo *et al* 2020). Guan *et al* (2020) proposed an IIOT (Industrial Internet of Things)-assisted blockchain-based energy trading system. A blockchain-based P2P transactions framework for IIOT transaction settings was proposed by Hu and Li (2021).

Residential battery storage capacities are relatively more affordable than larger entities such as shared facility controllers (Tushar *et al* 2014), power distribution companies, data centres, and so on. Furthermore, larger storage requirements require

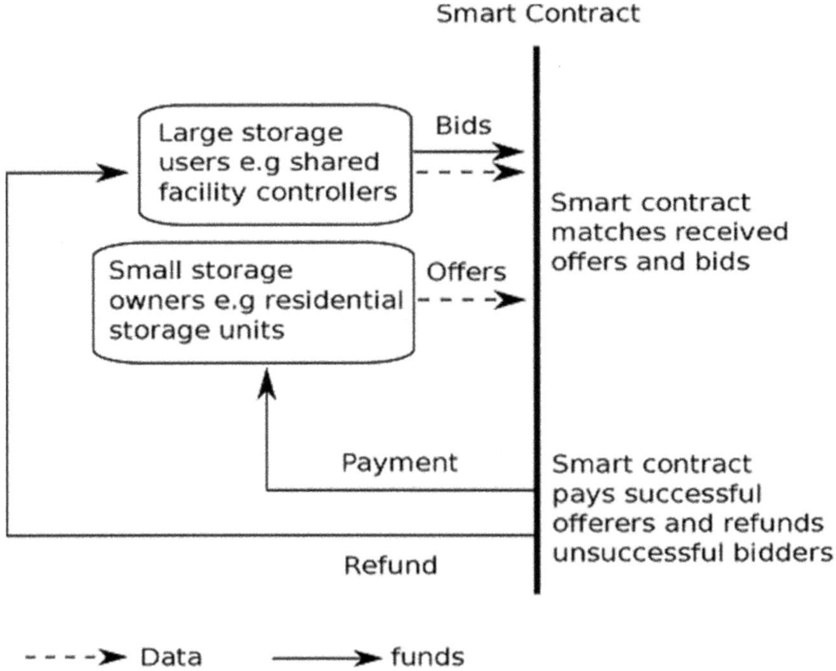

Figure 8.1. A block diagram representing a P2P storage capacity sharing scheme. Reproduced from Damisa and Nwulu (2022). CC BY 4.0.

more physical space allocation. Therefore, the pooling of multiple residential storage units offers potential benefits. Tushar *et al* (2016) investigated the sharing of residential storage facilities between individual facility owners and shared facility controllers. Finally, this chapter builds on the work presented in Damisa and Nwulu (2022).

8.2 Smart contract-based peer-to-peer energy storage-sharing architecture and algorithm

Figure 8.1 is a block diagram representing a blockchain-based energy storage-sharing scheme. Small-scale energy storage capacity owners (residential storage units in this case) send their offers to a smart contract, and large or utility-scale storage capacity utlilisers send their bids and funds to the smart contract. The contract uses its pre-programmed auction procedure to match buyers and sellers. Payments and refunds are made to the appropriate parties, where necessary. A real-world implementation of the scheme will require much more than a smart contract. Smart metres, desktop or mobile applications, cryptocurrency wallets, among others, will be needed.

8.3 Smart contract functions for the storage-sharing scheme

Constructor: This function gets executed at the point of contract deployment. In the case at hand, the address of the contract manager is assigned to a variable 'admin'.

ethValue: It allows the dollar equivalent of an ether to be set. The function restricts the privilege of setting this value to the 'admin'.

viewEthEquiv: With this function, an intending bidder can view the cost in ethers of the storage capacity being demanded.

RSUoffers: Residential storage units submit their storage capacity and price offers to this function.

vRSUoffers: The offers received in the foregoing function are ordered in ascending order based on price and returned to this function. The function checks to ensure that the returned data matches the original data received by 'RSUoffers' to ensure data integrity.

SFCbids: This function receives storage capacity and price bids, as well as funds sufficient to meet the required storage capacity.

vSFCbids: The bids received in the preceding function are ordered in a descending order based on price and returned to this function. To ensure data integrity, the function checks to ensure that the returned data match the original data received by 'SFCbids'.

matching: This is a crucial function, as it determines the winners of an auction round. It uses a pre-programmed auction procedure to match offerers and bidders and determines the price at which each pair transacts. Figure 8.2 shows the smart contract flow chart of the matching function.

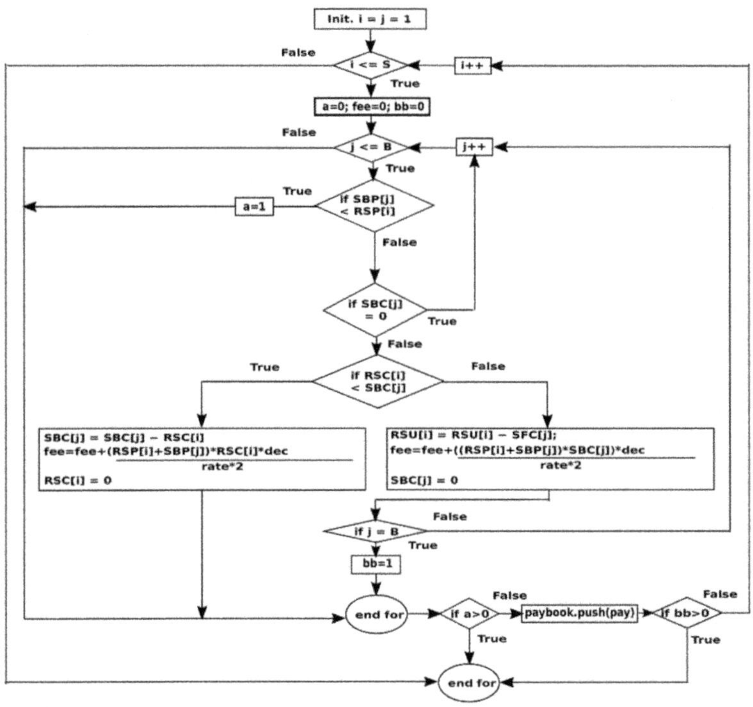

Figure 8.2. Flowchart of a smart contract matching function. Reproduced from Damisa and Nwulu (2022). CC BY 4.0.

settleRSU: This function is called by the 'admin' to instruct the smart contract to pay the successful sellers of storage capacity.
refundSFC: The 'admin' also calls this function to refund bidders who were unsuccessful during an auction round.

The P2P trading smart contract scheme proposed in Damisa and Nwulu (2022) is vulnerable when ordered offer data are being returned to the smart contract. While it ensures that the returned data matches the original data submitted by offerers, no measure has been put in place to guarantee that the data being returned are in the right order. Hence, an attacker can tamper with the order of the offers outside the smart contract to favour certain offerers, giving them a higher priority than other cheaper offers. This chapter introduces a new measure to check that the return offer data is rightly ordered.

8.4 Case study and discussion

In this study, we investigate the smart contract using three storage capacity market scenarios: a base scenario, price scenarios, and capacity scenarios. While the base scenario is obtained from the literature, the price and quantity scenarios are hypothetically constructed to represent extreme market conditions for storage capacity, testing the ability of the smart contract's ability to handle such cases.

8.4.1 Base scenario

See table 8.1 of Damisa and Nwulu (2022) for the base scenario data. RSUs and SFCs submit their offers and bids, respectively, to the smart contract. The offers are arranged from the lowest offer price to the highest, while the bids are arranged from the highest to the lowest price. Matching commences from the lowest offerer and the highest bidder, and the clearing price for each matched pair is the midpoint between the offer and bid prices. If, at any point the bid price falls below the offer price as matching progresses, then matching is terminated and the auction round is closed. The committed funds of unsuccessful bidders are refunded. Figures 8.3 and 8.4 show the smart contract's execution of the auction procedure, payments, and refunds.

Table 8.1. Sellers/buyers offer/bid amount and prices (offers> bids).

RSU group	Offer (\$ kW h^{-1})	Capacity (kWh)	SFC	Bid (\$ kW h^{-1})	Capacity (kWh)
A	50	160	a	40	300
B	55	470	b	35	350
C	45	450	c	44	230
D	50	140	d	40	480

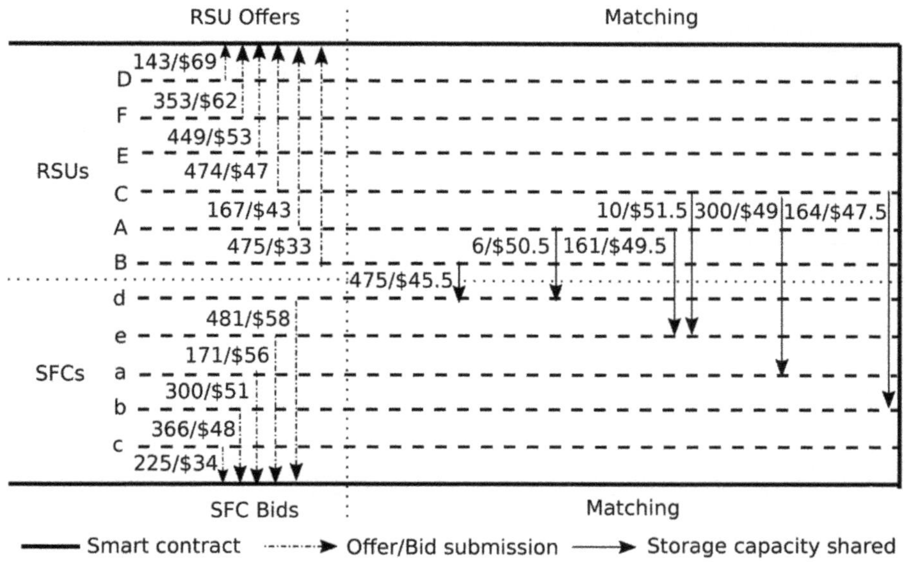

Figure 8.3. Smart contract execution of matching. Reproduced from Damisa and Nwulu (2022). CC BY 4.0.

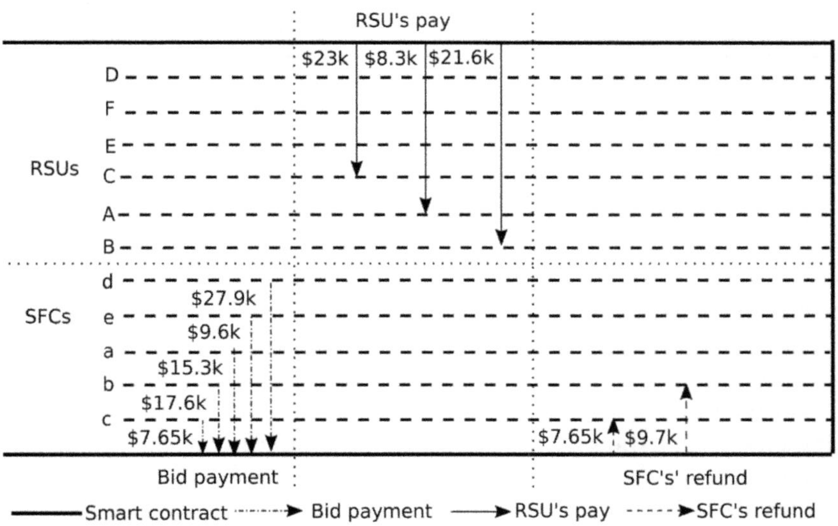

Figure 8.4. Smart contract's execution of payments and refunds. Reproduced from Damisa and Nwulu (2022). CC BY 4.0.

8.4.2 Price scenarios

The first case in the price scenarios investigates the response of the smart contract to cases where all the offer prices exceed the bids, i.e., the lowest offer price is greater than the highest bid price (table 8.1). In the second case, the lowest bid price exceeds the highest offer price (table 8.2). The price of Ether is assumed to be $3000.

Table 8.2. Sellers/buyers' offer/bid amount and prices (offers< bids).

RSU group	Offer ($ kW h^{-1})	Capacity (kWh)	SFC	Bid ($ kW h^{-1})	Capacity (kWh)
A	40	160	a	50	300
B	35	470	b	50	360
C	40	420	c	45	220
D	30	140	d	60	480

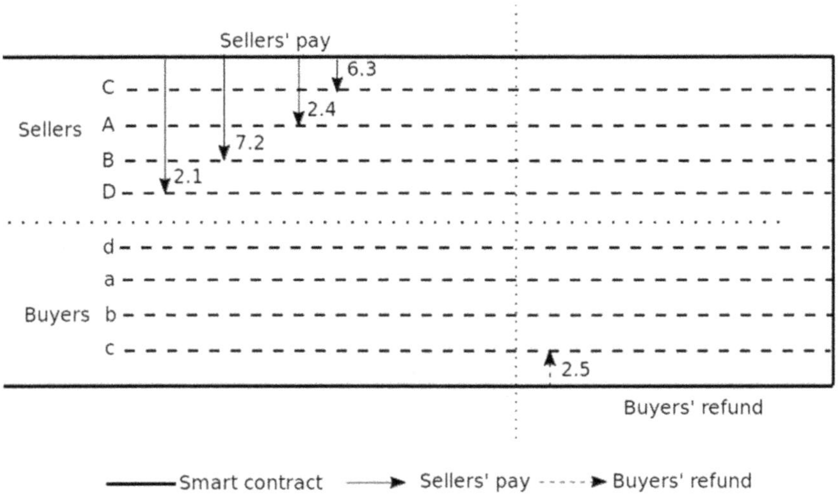

Figure 8.5. Payments and refunds by smart contract.

When the offers are all higher than the available bids, the fees spent by entities in the scheme are lost because no buyer or seller matches in the auction round. The smart contract administration fees generally include deploying the contract, setting the dollar equivalent of an Ether, verifying ordered offers/bids, matching buyers and sellers, and paying and refunding sellers and unsuccessful buyers, respectively. A potential seller pays a transaction fee to submit an offer to the smart contract, while a potential buyer is charged a transaction fee to submit a bid. Some or all of these fees can be lost in an auction round, such as the present one being considered. Mechanisms can be put in place to narrow the spread between offers and bids to avoid such a scenario. Larger clusters of potential sellers and buyers may also reduce the chances of realising this scenario. Figure 8.5 shows the payments made in Ethers by the smart contract to the appropriate parties.

8.4.3 Capacity scenarios

In the first scenario in the capacity scenarios the total amount of capacity demanded is less than what is available (table 8.3). The second assumes that capacity supply exceeds demand (table 8.4).

Table 8.3. Sellers/buyers' offer/bid amount and prices (demand< supply).

RSU group	Offer ($ kW h^{-1})	Capacity (kWh)	SFC	Bid ($ kW h^{-1})	Capacity (kWh)
A	40	360	a	50	100
B	35	470	b	45	160
C	50	480	c	35	220
D	70	240	d	60	280

Table 8.4. Sellers/buyers' offer/bid amount and prices (demand> supply).

RSU group	Offer ($ kW h^{-1})	Capacity (kWh)	SFC	Bid ($ kW h^{-1})	Capacity (kWh)
A	40	160	a	55	300
B	30	275	b	40	350
C	45	270	c	45	220
D	55	140	d	60	480

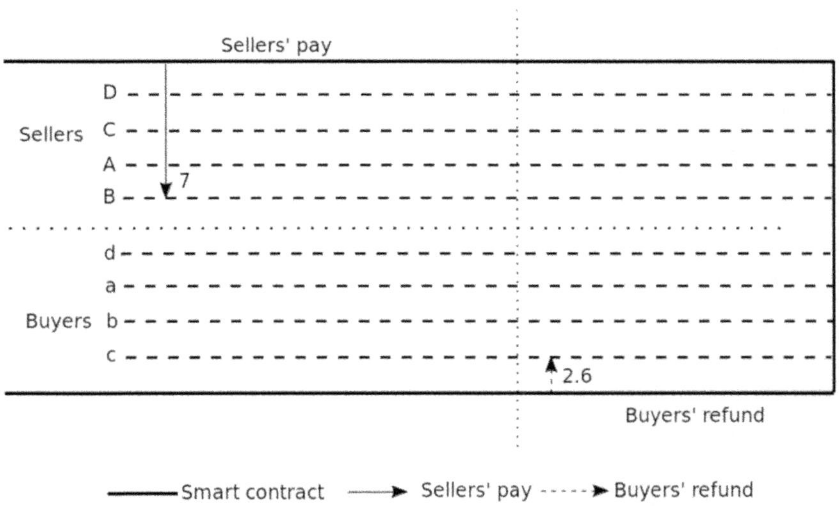

Figure 8.6. Payments and refunds by smart contract.

In quantity scenarios the price remains the main consideration when matching sellers and buyers. Regardless of the quantities available or demanded, offer and bid prices determine how matching is performed. Matching is impossible if a bid price is lower than an offer price. Consequently, there could be storage capacity supply and demand that is not matched due to price. The need for mechanisms that narrow the spread between offers and bids is further underscored. Figures 8.6 and 8.7 show the payments made in Ethers by the smart contract to the appropriate parties.

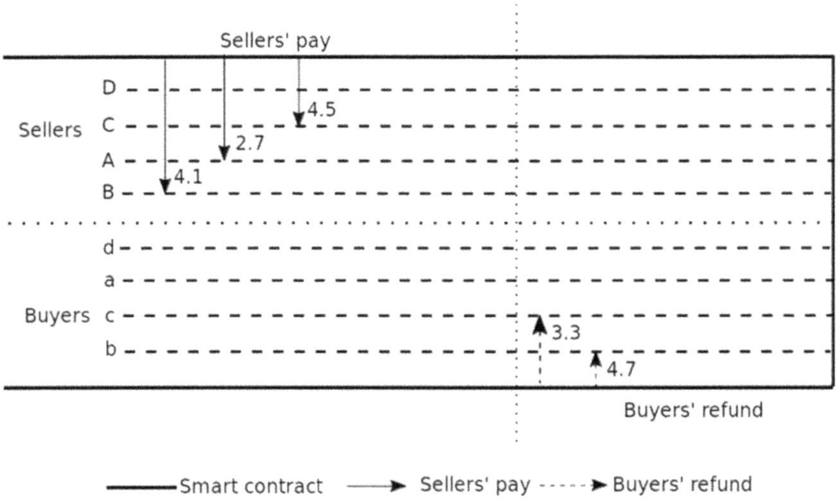

Figure 8.7. Payments and refunds by smart contract.

8.5 Summary

Pooling small-scale storage resources to serve large storage capacity needs improves the utility of the individual storage units, and offers cost and space savings to large-scale utilities. The transparency and trustlessness offered by blockchain-based solutions in such storage-sharing schemes can potentially drive participation of additional storage resources, thereby providing more capacity for large-scale utility. Significant work needs to be done to improve current work to achieve a practical blockchain-based storage-sharing scheme. In this work, a smart contract has been used to manually facilitate an auction-based procedure and effect payments. A practical system will involve data collection by third-party devices such as smart meters. In addition, user interaction with the smart contract will use a more user-friendly interface.

References

Abdella J and Shuaib K 2018 Peer to peer distributed energy trading in smart grids: a survey *Energies* **11** 1560

Chakraborty P, Baeyens E, Poolla K, Khargonekar P P and Varaiya P 2018 Sharing storage in a smart grid: a coalitional game approach *IEEE Trans. Smart Grid* **10** 4379–90

Dai R, Esmaeilbeigi R and Charkhgard H 2021 The utilization of shared energy storage in energy systems: a comprehensive review *IEEE Trans. Smart Grid* **12** 3163–74

Damisa U and Nwulu N I 2022 Blockchain-based auctioning for energy storage sharing in a smart community *Energies* **15** p.1954

Foti M, Mavromatis C and Vavalis M 2021 Decentralized blockchain-based consensus for optimal power flow solutions *Appl. Energy* **283** 116100

Guan Z, Lu X, Wang N, Wu J, Du X and Guizani M 2020 Towards secure and efficient energy trading in IIoT-enabled energy internet: a blockchain approach *Future Gener. Comput. Syst.* **110** 686–95

Guo J, Ding X and Wu W 2020 Combined cooling, heating, and power system in blockchain-enabled energy management arXiv preprint arXiv:2003.13416

Han D, Zhang C, Ping J and Yan Z 2020 Smart contract architecture for decentralized energy trading and management based on blockchains *Energy* **199** 117417

Hu W and Li H 2021 A blockchain-based secure transaction model for distributed energy in industrial Internet of Things *Alexandria Eng. J.* **60** 491–500

Li Y, Yang W, He P, Chen C and Wang X 2019 Design and management of a distributed hybrid energy system through smart contract and blockchain *Appl. Energy* **248** 390–405

Luo B, Shen X and Ping J 2020 Energy storage sharing mechanism based on blockchain *2020 IEEE 3rd Student Conf. on Electrical Machines and Systems (SCEMS)* pp 913–7

Münsing E, Mather J and Moura S 2017 Blockchains for decentralized optimization of energy resources in microgrid networks *2017 IEEE Conf. on Control Technology and Applications (CCTA)* pp 2164–71

Nunna H K, Sesetti A, Rathore A K and Doolla S 2020 Multiagent-based energy trading platform for energy storage systems in distribution systems with interconnected microgrids *IEEE Trans. Ind. Appl.* **56** 3207–17

Rahbar K, Moghadam M R V, Panda S K and Reindl T 2016 Shared energy storage management for renewable energy integration in smart grid *2016 IEEE Power & Energy Society Innovative Smart Grid Technologies Conf. (ISGT)* pp 1–5

Saini V K, Gupta V, Kumar R, Panigrahi B K and Mahmud M A 2020 Cloud energy storage systems for consumers and prosumers in residential microgrids *2020 IEEE Int. Conf. on Power Electronics, Drives and Energy Systems (PEDES)* pp 1–6

Sim J, Kim M, Kim D and Kim H 2021 Cloud energy storage system operation with capacity P2P transaction *Energies* **14** 339

Tushar W, Chai B, Yuen C, Huang S, Smith D B, Poor H V and Yang Z 2016 Energy storage sharing in smart grid: a modified auction-based approach *IEEE Trans. Smart Grid* **7** 1462–75

Tushar W, Chai B, Yuen C, Smith D B, Wood K L, Yang Z and Poor H V 2014 Three-party energy management with distributed energy resources in smart grid *IEEE Trans. Ind. Electron.* **62** 2487–98

van Leeuwen G, AlSkaif T, Gibescu M and van Sark W 2020 An integrated blockchain-based energy management platform with bilateral trading for microgrid communities *Appl. Energy* **263** 114613

Wang N, Zhou X, Lu X, Guan Z, Wu L, Du X and Guizani M 2019 When energy trading meets blockchain in electrical power system: the state of the art *Appl. Sci.* **9** 1561

Wang Y, Qiu W, Dong L, Zhou W, Pei Y, Yang L, Nian H and Lin Z 2020 Proxy signature-based management model of sharing energy storage in blockchain environment *Appl. Sci.* **10** 7502

Wang Z, Gu C, Li F, Bale P and Sun H 2013 Active demand response using shared energy storage for household energy management *IEEE Trans. Smart Grid* **4** 1888–97

Xie K, Zhong W, Li W and Zhu Y 2019 Distributed capacity allocation of shared energy storage using online convex optimization *Energies* **12** 1642

Zhao D, Wang H, Huang J and Lin X 2019 Virtual energy storage sharing and capacity allocation *IEEE Trans. Smart Grid* **11** 1112–23

Zhong W, Xie K, Liu Y, Yang C and Xie S 2020 Multi-resource allocation of shared energy storage: a distributed combinatorial auction approach *IEEE Trans. Smart Grid* **11** 4105–15

IOP Publishing

Blockchain-based Peer-to-Peer Transactions in Energy Systems

Nnamdi Nwulu and Uyikumhe Damisa

Chapter 9

Smart contract for peer-to-peer sharing of an EV charging station

The drive for a cleaner environment has contributed to the shift from combustion engine-based vehicles to electric vehicles (EVs). However, the limited range of EVs, coupled with the shortfall in EV charging stations, has stalled the adoption of EVs. An efficient EV charging station sharing system can improve this situation. In this chapter, a smart contract is conceptualised to facilitate the sharing of EV charging stations. Different tests performed on its logic will help to determine its accuracy. While the contract shows potential, further work needs to be done to make it useful in a practical setting.

9.1 Introduction and literature review

The global drive for cleaner energy use is leading the development and adoption of EVs. While conventional vehicles rely heavily on energy from fossil fuels, EVs consume electricity, which can be sourced from renewable energy sources. As a result, EVs are crucial in the fight against global warming. EVs can also contribute to power grid stability by providing additional services. In addition, they can serve as flexible demand for demand response initiatives. Power market operation and power system stability are greatly affected by EVs' charging and discharging capabilities (Huang *et al* 2021). EVs are integral to smart grid energy management (Khalid *et al* 2021). However, the global adoption of EVs is limited by a number of factors. The limitation of the driving range of EVs is a crucial factor preventing their adoption (Xia *et al* 2020). The difficulty of charging EVs also inhibits their development (Wang 2022). In addition, land scarcity and high construction costs have contributed to the slow development of EV charging facilities (Fu *et al* 2021). A dense and widespread EV charging infrastructure will be needed to support the growing adoption of EVs (Huang *et al* 2018). Private charging piles can be made available to the public through a secure and transparent peer-to-peer (P2P) system to improve the availability of EV charging infrastructure. This will improve the utilisation rate of these charging piles and promote the development of more private charging piles, which will drive the growth of the EV industry. Blockchain has proven to be effective in facilitating transparent and secure P2P transactions in a trustless environment. The use of blockchain technology in different parts of the EV value chain has been explored in the literature. To ensure that billing information is tamper proof, Jeong *et al* (2018) proposed a blockchain-based billing system for EVs and charging stations. In the scheme, billing information is stored on a blockchain after mutual authentication by an EV and a charging station. Gorenflo *et al* (2019) introduced the blockchain to provide immutability of data and shared data access to a legacy EV charging service provider setup, thus mitigating trust issues. Huang *et al* (2018) proposed a blockchain-based decentralised security model to enhance the security of transactions between charging stations and EVs. Fu *et al* (2020) used a consortium blockchain in an EV charging system, which improves the transparency and reliability of the EV charging information. Augello *et al* (2023) used the blockchain to share information between stakeholders in vehicle-to-grid programmes, which can be used to monitor the EV's health status and facilitate compensation.

This work focuses on sharing private electric vehicle charging stations using a blockchain-enabled system, which has also been explored in the literature. Kirpes and Becker (2018) investigated the storage and validation of EV charging data on a blockchain in a P2P EV charging station scheme, as well as the processing of payments for EV charging in a blockchain-based information system. Fu *et al* (2021) proposed a blockchain-assisted private charging station sharing scheme to enable the direct sharing of charging stations with EV owners. He *et al* (2021) developed a blockchain-based architecture for charging pile sharing involving entities such as EV users, private charging facility owners and operators, which offers secure computing, storage, and incentives. The blockchain-based platform proposed in Okwuibe *et al* (2020) connects EV users and charging stations, while managing EV charging requests to ensure that they are met within the capacity limit of the distribution grid. Finally, Pustišek *et al* (2016) introduced a blockchain-based concept in which EVs can autonomously select charging stations based on price and other possible factors, such as estimated charging duration and waiting time.

9.2 Smart contract-based peer-to-peer electric vehicle charging station sharing and algorithm

Figure 9.1 shows a block diagram representation of a smart contract-based P2P EV charger-sharing scheme. The EV owner requests offers from nearby charging stations, which are sent to the smart contract. The smart contract selects the most economical charging option based on their distances and price offers. The EV then pays for the amount of charging energy required to the smart contract. Following complete charging of the EV, the smart contract pays for the EV charger. In addition to a smart contract, a practical implementation of this scheme would

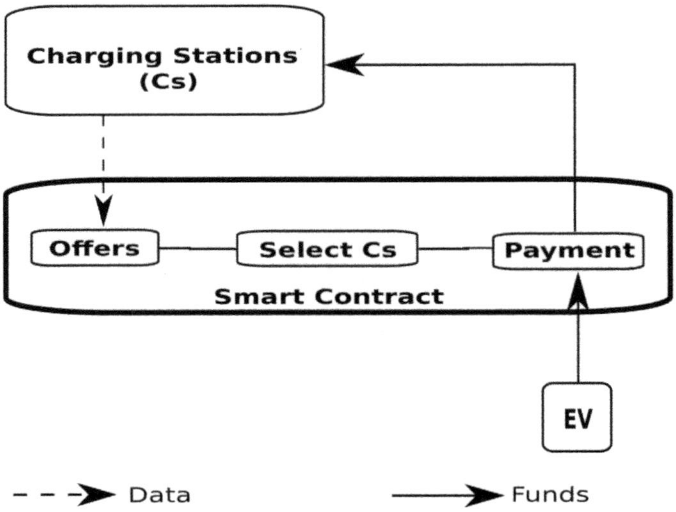

Figure 9.1. A block diagram representing an EV charging station sharing scheme.

involve external services such as a global positioning system to determine the position of charging stations relative to the electric vehicle and smart meters to capture the amount of energy spent charging. Oracles will also be needed to feed the smart contract with the equivalent value of ether in dollars.

9.3 Smart contract function for the electric vehicle charging station sharing scheme

The constructor is automatically executed when the smart contract is deployed. In our context, its execution involves assigning the deployer's address to a variable. This can be used to assign administrative privileges to the contract deploying company.

The function setRate can only be called by the contract deployer to set the value of the Ether in dollars.

The Function EVoffers offers the distance and price for each nearby EV charging facility. By evaluating these parameters, it selects the cheapest option. The distance measures how far the EV is from the charging station, and the price is the cost per unit of charging energy. In a practical setting, this distance can be obtained by integrating the Application Programming Interface (API) of third-party software.

The function ethValoffer indicates the amount in Ethers to be paid by the EV based on the cheapest offer selected by the foregoing function.

The pay4EvChing is a payable function that receives the indicated amount of ethers from the EV. The smart contract is an escrow account that holds funds until charging is complete and then pays the charger.

Function payEvCharger is called by the contract administrator to execute payment for EV charging to the EV charger.

9.4 Case scenarios and discussion

EVs are assumed to consume 0.15 kWh km^{-1} on average and the average cost of electricity is 20 cents kW h^{-1}. One Ether is assumed to be equivalent to $100. The EV charging offers received from charging stations are ranked based on their distance from the EV and their offer price. A charging demand of 30 kWh is assumed for all the scenarios investigated.

9.4.1 Base scenario

The base scenario is a random combination of EV charging station and offer price. As shown in table 9.1, charging offers are received from four charging stations. The smart contract selects the most economical option based on its distance from the EV and its offer price. In the present case, CS D is selected as expected. Consequently, the EV is required to pay for its 30 kWh demand at 20 cents kW h^{-1}, which is equivalent to $6. This amounts to 0.06 ether when converted at the rate of 1 ether to $100. The smart contract holds the funds until charging is over, after which the money is deposited into the EV charger D's account.

9.4.2 Scenario 1: all charging stations equidistant from electric vehicle

This scenario assumes that all the EV charging stations are equidistant from the EV (see table 9.2). As expected, since all charging stations are equally distant from the EV, the most economic offer can be selected based on the offer price alone. The smart contract's choice of charging station is CS B, which offers the lowest price.

Table 9.1. Charging station offer and distance.

Charging station	Cs offer (cents kW h^{-1})	Cs distance (km)
A	30	2
B	20	3
C	40	1
D	20	2

Table 9.2. Charging station offer and distance.

Charging station	Cs offer (cents kW h^{-1})	Cs distance (km)
A	30	3
B	20	3
C	40	3
D	25	3

Table 9.3. Charging station offer and distance.

Charging station	Cs offer (cents kW h^{-1})	Cs distance (km)
A	20	2
B	20	3
C	20	1
D	20	2

Again, the EV is required to pay for its 30 kWh demand at 20 cents kW h^{-1}, which amounts to \$6. This amounts to 0.06 ether when converted at the rate of 1 ether to \$100. The smart contract holds the funds until charging is over, after which the money is paid into EV charger B's account.

9.4.3 Scenario 2: equal offer price from all charging stations

The second scenario assumes that all the EV charging stations offer the same price (see table 9.3). Since all CS are offering the same charging price, the distance alone can be used to select the most economical charging option. In this case, the smart contract selects CS C, and deductions and payments are made accordingly.

9.5 Summary

Although the EV industry is gaining attention, a lack of sufficient charging stations may stall its development. An effective blockchain-based P2P EV charging station sharing scheme can potentially reduce range anxiety in EV owners, thus improving the adoption of EVs. The proposed smart contract has been used to demonstrate a fair selection of the most economical option for an EV based on its distance from a charging station and the cost per unit of energy offered. It is also an escrow account that receives and holds funds until the transacting parties complete their transactions. It is, however, only a step towards a practical system, which would need real time location data from external sources such as global positioning systems, charging data from smart meters, and ether value data from oracles.

References

Augello A, Gallo P, Sanseverino E R, Sciabica G and Sciumè G 2023 Certifying battery usage for V2G and second life with a blockchain-based framework *Comput. Netw.* **222** 109558

Fu Z, Dong P and Ju Y 2020 An intelligent electric vehicle charging system for new energy companies based on consortium blockchain *J. Clean. Prod.* **261** 121219

Fu Z, Dong P, Li S, Ju Y and Liu H 2021 How blockchain renovate the electric vehicle charging services in the urban area? A case study of Shanghai, China *J. Clean. Prod.* **315** 128172

Gorenflo C, Golab L and Keshav S 2019 Mitigating trust issues in electric vehicle charging using a blockchain *Proc. of the Tenth ACM Int. Conf. on Future Energy Systems* pp 160–4

He Y, Zhang C, Wu B, Geng Z, Xiao K and Li H 2021 A trusted architecture for EV shared charging based on blockchain technology *High-Confid. Comput.* **1** 100001

Huang X, Xu C, Wang P and Liu H 2018 LNSC: a security model for electric vehicle and charging pile management based on blockchain ecosystem *IEEE Access* **6** 13565–74

Huang Z, Li Z, Lai C S, Zhao Z, Wu X, Li X, Tong N and Lai L L 2021 A novel power market mechanism based on blockchain for electric vehicle charging stations *Electronics* **10** 307

Jeong S, Dao N-N, Lee Y, Lee C and Cho S 2018 Blockchain based billing system for electric vehicle and charging station *2018 10th Int. Conf. on Ubiquitous and Future Networks (ICUFN)* pp 308–10

Khalid R, Malik M W, Alghamdi T A and Javaid N 2021 A consortium blockchain based energy trading scheme for Electric Vehicles in smart cities *J. Inform. Secur. Appl.* **63** 102998

Kirpes B and Becker C 2018 *Processing Electric Vehicle Charging Transactions in a Blockchain-Based Information System* (AISeL)

Okwuibe G C, Li Z, Brenner T and Langniss O 2020 A blockchain based electric vehicle smart charging system with flexibility *IFAC-PapersOnLine* **53** 13557–61

Pustišek M, Kos A and Sedlar U 2016 Blockchain based autonomous selection of electric vehicle charging station *2016 Int. Conf. on Identification, Information and Knowledge in the Internet of Things (IIKI)* pp 217–22

Wang J 2022 A novel electric vehicle charging chain design based on blockchain technology *Energy Rep.* **8** 785–93

Xia S, Lin F, Chen Z, Tang C, Ma Y and Yu X 2020 A Bayesian game based vehicle-to-vehicle electricity trading scheme for blockchain-enabled internet of vehicles *IEEE Trans. Veh. Technol.* **69** 6856–68

Chapter 10

Blockchain in energy systems: conclusions and prognosis

Blockchain technology is poised to transform many sectors, offering unprecedented transparency, security, and decentralised transaction capabilities. However, its adoption is contingent on developing comprehensive and adaptive regulatory frameworks. Such structures should not only promote innovation but also ensure stakeholder protection and compliance with international standards. This chapter underscores the importance of collaborative policymaking involving governments, industry leaders, and academia. A holistic and informed approach to blockchain regulation can optimise its benefits, address potential challenges, and ensure its harmonious integration into the global socioeconomic landscape.

10.1 Introduction

The rapid rise of blockchain technology has captured the imagination of visionaries, sceptics, investors, and innovators alike. Its transformative potential extends beyond the realm of currencies, inciting palpable excitement across many industries, from finance to supply chain and healthcare to entertainment. This story began in 2009, when an entity or individual under the pseudonym Satoshi Nakamoto released Bitcoin, a decentralised digital currency system (Sasikumar *et al* 2022). The underpinning of this groundbreaking innovation was blockchain technology, which at the time was largely synonymous with cryptocurrencies.

However, as the world became more developed, the expansive utility of blockchain began to unfold. Its foundational principles hinted at broader applications that could redefine how businesses and societies function. Central to its appeal is the employment of cryptographic hashing, which ensures data integrity. Once inscribed on the blockchain, the information becomes virtually tamper-proof, creating a fortress against unauthorised alterations. This distributed ledger

doi:10.1088/978-0-7503-6295-5ch10

technology offers unmatched transparency, with every network participant accessing the same unaltered data, establishing both accountability and a consistent version of events. This inherent nature of the blockchain not only simplifies the auditing process but also introduces a systematic approach to regulatory compliance, reducing opportunities for fraudulent activities. Efficiency is another hallmark because as technology, by design, bypasses intermediaries, thus ensuring more direct transactions. The concept of smart contracts has further accentuated this efficiency, automating various processes and significantly diminishing friction points (Hader *et al* 2022). Another transformative aspect is the empowerment that it offers users over their own data, pointing towards a world where decentralised identity solutions could address prevailing privacy concerns and minimise data breaches.

On a global scale, blockchain holds the promise of democratising financial services. It can make cross-border transactions more affordable and extend services to the financially marginalised sectors of society. Its peer-to-peer essence inherently eliminates the need for intermediaries, reducing transaction fees, and bolstering overall efficiency. Beyond the digital realm, blockchain has the potential to validate the authenticity of sustainable goods, monitor carbon offsets, and counter environmental deceit. Furthermore, it provides a secure, timestamped record for intellectual property rights, creating a robust barrier against copyright infringements and piracy in our digital age (Damisa and Nwulu 2022). Blockchain technology can revolutionize the energy sector by automating and simplifying the energy trading process, increasing transparency and efficiency. This could lower costs and increase market accessibility for buyers and sellers. Blockchain technology can also improve the optimal management of energy systems, enabling peer-to-peer energy trade, reducing grid stress and increasing dependability. It can also aid in the increased integration of renewable energy sources, such as microgrids, by creating autonomous power systems that monitor the origin of renewable energy credits. Furthermore, blockchain technology can serve as the backbone of transparent trading platforms, accelerating the transition to a clean energy economy and reducing greenhouse gas emissions. As we navigate this dynamic landscape, it is evident that blockchain is not merely a technological marvel but a beacon of hope for a transparent, decentralised, and just future.

10.2 Limitations

Although the transformative nature of blockchain technology is undeniable, it is not without its challenges. For example, the architecture of many blockchains in their early stages had difficulty managing high volumes of transactions. This trade-off between decentralisation and scalability sometimes resulted in network congestion, leading to slow transaction times and hefty fees during peak usage. This was particularly evident during certain crypto-market surges when networks such as Ethereum faced overwhelming transaction costs. However, research into sharding and Layer 2 solutions offers some hope for the future. Another pressing concern has been the energy consumption associated with blockchain mining, especially in the Proof-of-Work (PoW) model (Alqarni *et al* 2023). Some critiques even equate Bitcoin's energy consumption to that of entire nations. Nevertheless, the horizon holds promise with the move towards more environmentally-friendly Proof-of-Stake (PoS) mechanisms.

The diverse world of blockchain platforms, each with unique protocols, often finds communicating and sharing data challenging. This fragmentation has sometimes made cross-platform transactions cumbersome. However, the industry is seeing promising endeavours like Polkadot and Cosmos projects that aim to introduce interoperability standards. On the regulatory front, the global acceptance of blockchain is uneven. Differing regulations across countries can hamper blockchain's adoption, with some initiatives even folding due to unclear regulatory landscapes (Kumari and Devi 2022). There is a growing realisation of the need for harmonised regulations that foster innovation while ensuring security.

While celebrated for their transparency, public blockchains also raise concerns about data privacy. Incidents of inadvertent exposure of sensitive data have highlighted this dilemma. Work is being done with privacy-centric blockchains and advanced technologies such as zero-knowledge proofs. For the average user, the blockchain can sometimes seem daunting. Challenges range from understanding how to manage private keys to navigating complex transactions. The industry has taken note, and there is a push towards more intuitive interfaces, tools for decentralised identity, and robust educational platforms. Jurisdictional challenges have also emerged, especially given blockchain's global nature (Lutfiani *et al* 2022). With data scattered across countries, potential legal conflicts can arise. Solutions might include geolocation-based data restrictions or hybrid strategies combining decentralisation with some centralised elements.

As blockchains grow, increased latency has also become an issue. Ethereum, for example, has faced challenges with growing data, leading to slow synchronisation. Innovative solutions such as data pruning and sharding are in the works to address these concerns. The storage of large amounts of data on blockchains poses its own set of challenges, especially given the decentralised nature of the technology. However, innovations such as the InterPlanetary File System (IPFS) offer decentralised storage solutions that might just be the answer. On the development side, crafting secure smart contracts is an intricate affair. Past vulnerabilities, such as the DAO hack, have emphasised the need for comprehensive audits and formal verifications (Taherdoost 2022). The community is actively working to refine smart contract tools and development processes.

Finally, the inherent complexity of blockchain cannot be overlooked, which requires a detailed understanding of cryptography, consensus algorithms, and specialised programming. The community recognises this steep learning curve, and is championing educational resources and tools to simplify the process for budding developers. While these challenges are significant, blockchain's journey showcases its resilience and adaptability, with innovative solutions continuously emerging to address its limitations.

10.3 Barriers

The transformative potential of blockchain technology has its share of hurdles. A key barrier is the limited knowledge and understanding of blockchain by the general population. Its complex technicalities can deter people and organisations from exploring its potential. There have been instances where misconceptions about blockchain have bred scepticism or resistance. Bridging this knowledge gap requires educational initiatives, workshops, and awareness campaigns. Blockchain's technical aspects present another challenge. Issues such as managing nodes, wallets, and private keys can be daunting. Cases where users have struggled with transactions or even faced security breaches serve to underline these complexities. However, efforts are underway to simplify user interfaces, create user-friendly wallets, and develop comprehensive tutorials. Marrying blockchain with existing legacy systems is not straightforward. Compatibility issues emerge when trying to blend blockchain with traditional processes (Geleziunaite and Manion 2023). The challenges faced by organisations in shifting to blockchain-based solutions underscore this problem. Interoperability standards and gradual migration strategies might be the solution.

Another thorn in blockchain's side is uncertain or inconsistent regulatory frameworks. Varying regulations can hinder blockchain's global adoption and projects have been delayed or shelved due to regulatory ambiguities. Lobbying efforts, regulatory sandboxes, and international cooperation aim to introduce clear and adaptive regulations. Scalability is another notable challenge. High transaction volumes can lead to network congestion, resulting in slow transaction confirmations and high fees, which affect usability. Solutions such as Layer 2 scaling, sharding, and advancements in consensus algorithms are being explored to combat this issue. While beneficial, transparent blockchains raise concerns about data privacy and security. Incidents of data exposure due to blockchain's transparency have occurred (Lennerfors and Murata 2023). However, privacy-centric blockchains, zero-knowledge proofs, and cryptographic techniques promise enhanced data privacy.

Adopting blockchain requires investment, and the costs can be a deterrent. High initial expenses can make businesses hesitate. However, cloud-based solutions, consortium models, and gradual adoption might cut these costs. Interoperability is a significant challenge. The absence of standardised protocols across blockchain platforms makes integrating them problematic. Initiatives such as cross-chain bridges and consortiums working on common standards might provide a solution. Environmental concerns, especially around energy-intensive blockchain mining, cannot be ignored. Criticisms of the energy-consuming PoW mechanisms have arisen (Andrei and Treiblmaier 2023). However, there is a shift towards more environmentally-friendly mechanisms such as PoS.

Decentralised networks bring their own governance challenges. There are examples where disagreements among stakeholders have led to network forks. On-chain governance and consensus-reaching mechanisms offer hope in this domain. Resistance to adopting blockchain due to its unfamiliarity is natural. However, successful case studies of early adopters provide motivation. The challenge also lies in finding professionals who are skilled in both blockchain and specific industries. The need for interdisciplinary expertise is paramount. Therefore, cross-disciplinary collaboration and training is gaining emphasis. The tokenisation of real-world assets introduces its own set of complexities, especially with regard to regulatory and legal aspects (Elsayed 2023). Despite this, regulatory frameworks such as Security Token Offerings (STOs) are paving the way for clearer tokenisation paths.

Infrastructure, especially in underdeveloped areas, remains a challenge. Projects aiming to improve connectivity and provide infrastructure in these areas are gaining traction. Intellectual property concerns arise, especially in open source development. Instances of disputes have been recorded. Licensing models and blockchain-based solutions might offer protection. Ensuring the security of smart contracts is vital. Coding vulnerabilities have resulted in exploited contracts and losses. Rigorous auditing and developer education are pivotal in addressing this problem. User interfaces that are not intuitive for the average person present another challenge. Efforts are made to design user-friendly interfaces and educate users. Furthermore, even private and consortium blockchains face scalability issues (Chen *et al* 2022). Solutions are being explored through hybrid architectures and optimised consensus algorithms.

The blockchain landscape is fragmented, with numerous platforms and protocols. This fragmentation can lead to confusion. Standardised protocols and interoperability frameworks are being developed to tackle this issue. The perception of blockchain as an overhyped technology has led to scepticism. However, focusing on realistic expectations and feasible use cases might change this perception. Finally, the lack of industry standards in areas such as smart contract languages and token standards can be problematic. Efforts by consortiums to establish universal standards are underway with the aim of seamless interoperability.

10.4 Policy and regulatory measures

Policy and regulatory measures are paramount in facilitating the adoption and integration of blockchain technology in various sectors. These measures serve as guiding principles, ensuring that innovation thrives while also safeguarding consumer interests, ensuring compliance, and upholding the integrity of the financial and tech ecosystems. A clear regulatory landscape promotes trust among blockchain developers and investors. However, ambiguous or contradictory regulations can deter investments and innovation. Regulations that are not clear-cut or inconsistent across jurisdictions can create barriers for blockchain projects. For instance, several businesses have relocated their operations due to unfriendly or unclear regulatory environments in their home countries (Sheelvanth *et al* n.d.). Regulators should provide explicit guidelines on digital assets and engage in international dialogues to ensure harmonisation of cross-border blockchain applications.

Regulatory sandboxes serve as testing grounds, enabling blockchain startups to experiment without the constraints of full-scale regulatory compliance. These controlled environments ensure that both innovation and oversight coexist. Countries such as Singapore and the UK have seen successful blockchain startups emerge from their sandbox initiatives. Clear criteria for sandbox participation can ensure a harmonious balance between innovation and regulation. Although blockchain promises transparency, it must be balanced with data protection regulations. Conflicts often arise between the inherent transparency and privacy laws such as the General Data Protection Regulation (GDPR). The cases of personal data leaks on public blockchains highlight the need for enhanced privacy measures. Regulations should emphasise privacy while also recognising blockchain's innate characteristics, pushing for innovations such as zero-knowledge proofs. Since they are self-executable code, smart contracts pose challenges to traditional legal enforcement. Ensuring legal recognition and a mechanism for enforcement is crucial. (Shinde *et al* 2022). Recognising smart contracts as binding and creating blockchain-native arbitration systems can help address these challenges.

The increasing number of blockchain platforms necessitates standardisation to ensure smooth interoperation. Fragmented blockchain ecosystems can prevent seamless data exchange and transactions. Cross-chain transactions often face hurdles due to differing standards across blockchain platforms. Promoting standardised protocols and supporting cross-chain initiatives can ensure seamless operations. As blockchain platforms and cryptocurrencies grow in popularity, protecting naive consumers and investors becomes paramount. The decentralised nature of blockchain creates opportunities for fraud and scams. Numerous ICOs have defrauded investors by providing misleading or simply false information. Strict guidelines for token offerings and investor education can mitigate such risks. Some blockchain mechanisms, especially PoW, have significant environmental implications. Blockchain energy consumption, especially in cryptocurrencies such as Bitcoin, has come under scrutiny (Motahhir and Bossoufi 2023). The energy consumption has surpassed that of some countries, raising environmental concerns. Incentivising energy-efficient consensus mechanisms can address these concerns.

Blockchain adoption in the energy sector faces three main obstacles: technological, financial, and legal. Technical challenges include scalability, complexity, and interoperability, as well as security and privacy concerns. Financial challenges include cost, lack of incentives, and lack of thorough regulation. Additionally, jurisdictional divisions can pose challenges for businesses deploying blockchain technology across multiple countries.

Regulatory impediments include the lack of thorough regulation, which puts companies at risk and causes uncertainty. Additionally, the lack of knowledge and understanding among stakeholders in the energy sector can hinder adoption. The conservative nature of the energy sector may make implementing cutting-edge technology like blockchain challenging.

Despite these challenges, there is widespread agreement that blockchain technology holds great promise for the future of the energy sector. As it develops and matures, it is expected to see a significant rise in application and adoption hurdles are removed. The energy sector is by nature conservative, making it difficult to implement cutting-edge technology like blockchain. Despite these obstacles, blockchain technology holds great promise for the future of the energy sector.

Regular policy evaluations ensure that regulations evolve alongside technological advancements. Blockchain's evolving nature necessitates adaptive policies. Continuous dialogues, policy reviews, and international collaboration can help to align regulations with technological progress. The global nature necessitates consistent regulations across borders. Inconsistent international regulations can hinder cross-border blockchain projects. Some blockchain projects face challenges when extending services internationally due to regulatory discrepancies. A synchronised global approach can facilitate cross-border blockchain initiatives. Policy measures play a pivotal role in promoting research and innovation on blockchain. Funding and resources are crucial for long-term R&D in blockchain technology. Several promising blockchain initiatives have stagnated due to a lack of resources or funding (Hader *et al* 2022). Governments can offer grants, establish innovation centres, and provide incentives to encourage R&D.

Collaborations between the government and the private sector can provide a balanced approach to blockchain regulations. Industry insights are crucial to effective regulatory frameworks. Successful blockchain policies often emerge from close collaborations between governments and industry experts. Establishing advisory boards and facilitating open dialogues can help in co-creating effective regulations. Educating policymakers is crucial to making informed regulatory decisions. Misconceptions can lead to suboptimal or restrictive policies. Some regulatory decisions have faced criticism due to apparent misunderstandings of blockchain technology. Regular workshops, collaborations with academia, and training initiatives can help to ensure informed decision-making. In conclusion, as blockchain technology continues its path towards widespread adoption, it is imperative that policy and regulatory frameworks should evolve in tandem, addressing challenges while also fostering innovation. Blockchain's transformative potential can be fully realised through collaboration, continuous learning, and adaptability.

10.5 Summary

In an era where technological advancements are redefining industries, blockchain stands out as a transformative force with the potential to revolutionise transactions, trust mechanisms, and data transparency. There is significant potential for Blockchain Technology to revolutionize the Energy Industry. The technology has a significant role in advancing energy security and mitigating energy poverty. The technology is also an enabling technology towards achieving Sustainable Development Goal (SDG) 7 which seeks for affordable and clean energy. However, to harness its full potential and ensure its sustainable integration into our socioeconomic structures, a balanced and clear regulatory framework is essential. Such a framework should foster innovation, protect stakeholders, and adapt to the evolving nature of the technology. Collaborative efforts between governments, industry experts, academia, and the private sector are vital to crafting informed policies. As we move forward, our regulatory approach must align with the principles of inclusivity, transparency, and adaptability, ensuring that blockchain technology benefits society at large.

References

Alqarni M A, Alkatheiri M S, Chauhdary S H and Saleem S 2023 Use of blockchain-based smart contracts in logistics and supply chains *Electronics* **12** 1340

Andrei and Treiblmaier H 2023 Blockchain technology as a driver of economic development in small economies: a dynamic capabilities framework *J. Decision Syst* 1–26

Chen C, Zhang L, Li Y, Liao T, Zhao S, Zheng Z, Huang H and Wu J 2022 When Digital Economy Meets Web3.0: applications and challenges *IEEE Open J. Comput. Soc.* **3** 233–45

Elsayed A N 2023 *The Use of Blockchain Technology in Education: A Comprehensive Review and Future Prospects* (Social Science Research Network)

Geleziunaite G and Manion S 2023 'Block-Change': exploring change management principles to overcome challenges in blockchain adoption *J. Br. Blockchain Assoc.* **6** 1–7

Hader M, Tchoffa D, Mhamedi A E, Ghodous P, Dolgui A and Abouabdellah A 2022 Applying integrated Blockchain and Big Data technologies to improve supply chain traceability and information sharing in the textile sector *J. Ind. Inform. Integr.* **28** 100345

Kumari A and Devi N C 2022 The Impact of FinTech and blockchain technologies on banking and financial services *Technol. Innov. Manag. Rev.* **12** 1–11 https://timreview.ca/article/1481

Lennerfors T T and Murata K 2023 *Ethics and Sustainability in Digital Cultures* (Milton Park: Taylor and Francis)

Lutfiani N, Apriani D, Nabila E A and Juniar H L 2022 Academic certificate fraud detection system framework using blockchain technology *Blockchain Front. Technol.* **1** 55–64

Motahhir S and Bossoufi B 2023 *Digital Technologies and Applications: Proc. of ICDTA'23 (Fez, Morocco)* (Berlin: Springer)

Sasikumar A, Vairavasundaram S, Kotecha K, Indragandhi V, Ravi L, Selvachandran G and Abraham A 2022 Blockchain-based trust mechanism for digital twin empowered Industrial Internet of Things *Future Gener. Comput. Syst.* **141** 16–27

Sheelvanth L and Gundge Y V 2023 An implementation of blockchain technology on forensic evidence management system *Int. Res. J. Modern. Eng. Technol. Sci.* **5** 2582–5208

Shinde R, Patil S, Kotecha K, Potdar V, Selvachandran G and Abraham A 2022 Securing AI-based healthcare systems using blockchain technology: a state-of-the-art systematic literature review and future research directions *arXiv.org.*

Taherdoost H 2022 Blockchain technology and artificial intelligence together: a critical review on applications *Appl. Sci.* **12** 12948

Damisa U and Nwulu N I 2022 Blockchain-based auctioning for energy storage sharing in a smart community *Energies* **15** 1954

www.ingramcontent.com/pod-product-compliance
Ingram Content Group UK Ltd.
Pitfield, Milton Keynes, MK11 3LW, UK
UKHW051850100926
8556IPUK00005B/547